Ehsan Pashai

Development of Enzyme-Based Biosensor and Nano-biosensor

Ehsan Pashai

Development of Enzyme-Based Biosensor and Nano-biosensor

for Nitric Oxide (NO) Detection

Noor Publishing

Imprint
Any brand names and product names mentioned in this book are subject to trademark, brand or patent protection and are trademarks or registered trademarks of their respective holders. The use of brand names, product names, common names, trade names, product descriptions etc. even without a particular marking in this work is in no way to be construed to mean that such names may be regarded as unrestricted in respect of trademark and brand protection legislation and could thus be used by anyone.

Cover image: www.ingimage.com

Publisher:
Noor Publishing
is a trademark of
International Book Market Service Ltd., member of OmniScriptum Publishing Group
17 Meldrum Street, Beau Bassin 71504, Mauritius

Printed at: see last page
ISBN: 978-620-0-07486-7

Development of Enzyme-Based Biosensor and Nano-biosensor for Nitric Oxide (NO) Detection

First Edition

Ehsan Pashai[1]

1. PhD in Chemical Engineering, Babol Noshirvani University of Technology, Babol, Iran

Contents

Abstract

In this study, biosensors based on the cytochrome c oxidase enzyme (Nafion/Cyt c Oxidase/AuNPs-PANI-ErGO/Au) and cytochrome c protein (Nafion/Cyt c/AuNPs-CS-MPA/SAMs-Au) and a graphene-based sensor (Nafion/AuNPs-PANI-ErGO/Au) were fabricated for NO detection and were evaluated by voltammetry and Chronoamperometry methods. Gold electrode was used for all three sensors. In all of these sensors, a Nafion membrane was used to prevent interferences. Several performance parameters such as concentration of enzyme, pH, and volume of Nafion, also the effects of interferences, repeatability, reproducibility and stability were evaluated. The structural, surface and electrochemical characteristics of the nanocomposite layers were investigated using FESEM and UV-vis tests. FESEM images confirmed the correct formation of these layers on the surface of the electrode. The results of the UV-vis test showed that the CS-AuNPS-MPA and AuNPS-PANI-ErGO nanocomposite layers had high electrical conductivity and created a suitable environment for catalytic activity of the enzyme and protein. Electrochemical and electro-catalytic performance of modified electrodes in the absence and presence of NO were evaluated by cyclic voltammetry and Chronoamperometry methods. The fabricated sensors based on enzyme, protein and graphene had a high sensitivity of 0.629 $\mu A.\mu M^{-1}$, 0.199 $\mu A.\mu M^{-1}$ and 0.113 $\mu A.\mu M^{-1}$ and detection limit of 0.016 μM, 0.045 μM and 0.25 μM, respectively. In addition, these biosensors respectively showed a short response time 4 s, 5 s and 4 s, and wide linear range of 0.1-35 μM, 10-215 μM and 0.8-86 μM. The low apparent Michaelis–Menten constant of 0.094 and 0.047 mM for enzyme and protein respectively, indicated that produced nanocomposites provided biocompatible environment for immobilized enzyme and proyein and assisted them to retain high bilogical activity. The obtained results indicating that these newly developed sensors were quite selective and stable which were used for accurate and exact detection of NO. The comparison between these three sensors showed that the biosensor based on Cyt c Oxidase enzyme had the best detection limit (0.016 μM) and the highest sensitivity (0.629 $\mu A.\mu M^{-1}$) and biosensor based on Cyt C protein had the widest range of linear responses (10-250 μM). In general, it can be concluded that the enzyme-based biosensor (Nafion / Cyt c Oxidase / AuNPS-PANI-ErGO / Au) showed the best performance.

Keywords: Biosensor, Cytochrome c oxidase enzyme, Cytochrome c protein, Graphene, Gold nanoparticles, Chitosan, Polyaniline, Cyclic voltammetry, Chronoamperometry.

CHAPTER

1

Introduction

1-1 Introduction

Development of quality of life is one of the most important research goals among human society. Naturally, quality of life depends on disease control, food, industrial operating conditions and environmental quality. In all these areas, rapid, simple, sensitive, and reagent-free methods are needed to control the key parameters. Applications of these rapid methods include veterinary, food, horticultural, pharmaceutical, petrochemical, environmental protection and defense. Conventional chemical detection methods, however slow and expensive, are non-specific and interfering. One of the quick detection tools is a biosensor. A biosensor is an analytical tool that can convert a biological response to an electrical signal to detect an Analyte or determine its concentration during a multi-step process. This biological response can be the result of a wide range of biological compounds such as enzymes, receptor ligand interaction, antibody antigen reaction, nucleic acid interaction, or even complete cell metabolism.

Depending on the type of biosensor and transducer used, biosensors have different types. If the diagnostic part of a sensor consists of an enzyme or a number of enzymes, a piece of living animal or plant tissue, or whole cells derived from microorganisms such as bacteria, fungi or molds, it is considered a catalyst biosensor.. Enzyme-based sensors measure the rate of reaction catalyzed by the enzyme as the basis of their reaction. Therefore, any signal related to the rate of this reaction can be used for identification [1]. Enzymes are proteins with high catalytic activity groups and are produced by living organisms. Enzymes are proteins with high catalytic activity groups and are produced by living organisms, but their unique ability to facilitate biochemical reactions without altering their structure and activity distinguishes them from proteins.

Enzymes act as catalysts in biochemical reactions and increase the rate of reaction by several folds and are biodegradable in addition to being specific and fast acting [2]. Enzymes can catalyze reactions in different states: as single molecules in the soluble mass with other substances or as attached to the surface. The concept of "enzyme fixation" refers to an "enzyme that is physically bound to a space by a specified substrate while maintaining catalytic activity" and it can be used several times. "Catalyst stabilization in many cases enhances the performance of industrial processes and improves economic costs [3].

The transducer transforms the response from the interactions between the analytic and the sensor element into a measurable and defined signal. The transducers are used as electrochemical, optical, thermal or as integrated sensors (using these transducers with magnetic and piezoelectric transducers). The biosensor selectivity is affected by the sensor element and its sensitivity to the transducer part.

Electrochemical sensors designed by integrating biological components and electrochemical transducers or modified with nanoparticles play an important role in biosensor studies. Over the past decade, advances in nanotechnology have led to the production of various nanomaterials that have various applications and play an important and influential role in the development and advancement of biosensors. Nanomaterials because of their unique properties are greatly improved catalytic and electrochemical activity of many proteins in the electrodes modified with these materials (including titanium dioxide, metal nanoparticles such as gold and silver, carbon nanotubes and the like). Therefore, the sensitivity and efficiency of biosensors are improved by using nanomaterials. Different types of nanostructures have been investigated as environments for protein fixation. In these studies, fixation was performed using techniques such as protein adsorption, covalent binding, protein encapsulation, and complex synthetic methods. Findings have shown that these materials not only provide a compatible substrate for the organization of protein molecules, but also enhance the electron transfer process between the protein molecule and the electrode.

Nitric oxide (NO) has been extensively studied because of its effects on air pollution and in particular biological systems. It is involved in many physiological and pathological processes. Abnormal NO production causes some important biological activities as well as many diseases. Deficiency or over-production of NO can lead to diseases such as hypertension, stroke, Parkinson's, Alzheimer's and diabetes (types I and II). Therefore, identification of NO is essential in biological activities. Since NO, a hydrophobic free radical, has a short half-life (less than 6 seconds); it is difficult to identify and requires rapid procedures. One of the quick identification methods is the use of biosensors.

Extensive studies have been conducted over the past decade to provide sensors or biosensors for the detection and detection of NO. Use of a novel combination of conductive polymers, nanomaterials and proteins or enzymes to detect NO, this research is one of the fascinating studies in the field of gas nanosensors in the world. In addition, no reports of sensors or biosensors for the detection of NO gas have been recorded in Iran.

1-2 Research purposes

The main purpose of this study was to develop graphene-based sensors and biosensors based on cytochrome c protein (Cyt c) and cytochrome c oxidase (Cyt c oxidase) to detect NO in a short time with high sensitivity and stability.

In this study, the surface of the gold electrode is prepared using nanocomposites containing gold nanoparticles, graphene and polymers such as chitosan and polyaniline and suitable for stabilization of the cytochrome c protein (Cyt c) and the cytochrome c oxidase enzyme. UV-vis and FESEM analyzes are used to characterize the nanocomposites and to investigate the electrode surface morphology. The electro-catalytic behavior of the electrodes made by cyclic voltammetric and chronoamprometric tests will be examined and the calibration diagram of the sensors constructed will be presented. Sensor performance parameters such as response time, detection limit, linear response range, stability, and selectivity will also be reported.

CHAPTER

2

Research background

2-1 Introduction

The first step in studying and researching any scientific field is to familiarize yourself with common definitions and terms and to carefully examine the results of research reported in that field. This chapter describes all the common definitions and terms used in sensors and biosensors. In addition, some studies in the same field as the present study are briefly mentioned.

2-2 Sensor

Sensors are tools for detecting and determining chemical and physical properties such as pressure, humidity, temperature, concentration of materials, etc. In recent years, developing industries require precise and rapid measurement methods, which can be achieved by the use of sensors [4].

The basis of the sensor work is that it converts physical and chemical changes into a measurable message, which is recorded by a particular instrument. A mercury thermometer is an example of a simple, primitive sensor. In a glass thermometer, an increase or decrease in temperature causes an expansion or contraction of the liquid inside it, which is measured by a graded glass tube. In the human body, some components also play the role of sensors. Tongue and nose naturally measure taste and smell. In the nose, when chemicals reach the olfactory bulb (a biological component), an electrical signal is generated and transmitted to the brain by the olfactory nerves. The brain eventually converts the produced signal into what is referred to as the sense of smell [5, 6].

2-3 Biosensor

A biosensor is a device in which a biodegradable component is connected to or integrated with a transducer. Classical biosensors are defined as analyzing devices in which a biological component is connected to a transducer and converts a biological signal into a measurable electrical signal.

In a biosensor, substances that are measured by a biosensor are called substrates. There are wide potential applications for different types of biosensors. Biosensors can be used to measure various materials. Such as carbohydrates (glucose, fructose, sucrose ...), alcohols (methanol, ethanol and other alcohols), fats (such as glycerol); amino acids (such as lysine and histidine), organic acids (e.g. Maleic acid and acetic acid), vitamins, toxins, heavy metals, microbes, bacteria, viruses, etc. [7].

Biosensor manufacturing utilizes the expertise of various disciplines such as biochemistry, immunology, optical physics, electrochemistry, materials science, semiconductors, electronics, and thousands of other science and engineering fields.

2-4 Biosensor Components

The biosensor components include the analyte, the biocatalyst, the transducer, the amplifier and the display.

2-4-1 Types of sensor elements

The element acting on the analyte in the biosensor is called the sensor element. In fact, it contains a portion of the biosensor that is sensitive to the analyte.

The importance of these components in their very specific performance towards the specific substrate is thereby preventing the interference of intrusive substances which may render many measurement methods ineffective. The biological component may catalyze the substrate reaction (such as the enzyme) or selectively bind to the substrate. The use of the enzyme as a biological component is more common than other materials.

Biosensors are the major determinants of biosensor selection. These elements are so strong that they only bind to a particular substrate and do not react with other substrates.

Enzyme

Enzymes are biomaterials that are of great use and may be used either in pure form or in microorganisms or in a portion of tissue. These materials are biocatalysts for specific reactions and can attach themselves to specific substrates. The performance of these materials in biosensor manufacturing is related to their catalytic performance [8].

Antibodies

Antibodies produced in organs are proteins that bind to hostile antigens, repel them and prevent damage to the body. Unknown antigens can be determined by labeled antibodies. Labeling can be accomplished by radioisotopes, fluorescent probes, or chemical luminescence probes or metal labels [9, 10]. Antibody-antigen binding is much stronger and more specific than the enzyme-linked substrate binding. In fact, they act very differently from one species to another. These materials are often used as markers. The advantages of this method include the high selectivity of these materials and their strong bonding. Lack of catalytic effect is one of their disadvantages.

Nucleic Acids

Nucleic acids also act like antibodies in different ways. The specific function between the open heads of two strands of nucleic acid is the origin of the genetic code that transcribes the characteristics of all parts of the living cell and thus inherits all the characteristics of each member of a species. Thus, for every molecule such as a protein or enzyme that is created within the cell, there is a nucleic acid that is genetically coded. DNA probes are useful for diagnosing genetic diseases, cancer, and viral infections [11].

Receptors

Most of the researches on receptors relate to neurotransmitters and the identification of transmitter, transmitter antagonists and neurotoxins. The use of this field as biosensors is still at the beginning [6].

Binding of a ligand to a receptor, it gives rise to a strong physiological response. Instead of reacting to a particular material, such biosensors tend to favor a group of compounds with structural similarities. Neurotransmitters and hormone receptors are biosensors of the body [9, 10].

2-4-2 Transducers

The chemical or physical change caused by the effect of the sensor element on the analyte is converted by a converter into a measurable (electrical) message.

Electrochemical transducers

The basic principle behind electrochemical biosensors is that chemical reactions produce or consume ions or electrons, which in turn causes some changes in the soluble electrical properties that it can be used for diagnosis and as a measurement scale [12].

Potentiometric

The voltage resulting from the biochemical reactions in a solution due to the presence of a particular biological molecule allows the current to pass through the solution and thus measure the potential for reduction or oxidation. Potentiometric is based on measuring the potential of a cell in zero current [13].

Voltammetry

The presence of catalytic enzymes in biochemical reactions produces electrophoretic biomolecules that are responsible for the current flowing, which in this way is the measured parameter. At voltammetry, an increase (or decrease) of potential is applied to the cell to oxidize (decrease) the material to be measured and a rapid increase (decrease) in the cell current is observed to create the maximum current. If the oxidation potential value is known, the potential can be directly adjusted to that value and the current recorded. This method is known as Amperometry [11]. Many biosensors work based on amperometry.

Conductimetric

It is a technique that measures changes in ion concentration. The parameter measured in conductivity is electrical conductivity and solute resistance. Solutions containing ions conduct electricity. Changes in the conductivity coefficient or specific resistance of the solution due to the discharge of ions and electrons into the solution occur as a result of the electrochemical reaction. The magnitude of this conductivity may be altered by chemical or biochemical reactions. The relationship between conductivity and concentration depends on the nature of the reaction. In the conductivity method, the measurement is very simple [11].

Ion-sensitive

The parameter measured in this method is the surface electric potential. In this method, miniaturized biosensors can be built using a Field-effect transistor (FET). Here the electrode is coated with polymers, which in the role of the membrane allow the passage of ions of a particular analyte. Therefore, the analyte ions alter the electrode potential of the electrode surface as well as the pH of the solution. Hence, the electrical potential of the surface is measured. These types of biosensors are known as ion-sensitive FETs [14].

Optical transducers

With the development of optical fibers, a new trend in life has emerged that has made it possible to flex and miniature. The methods used in optical biosensors include absorption spectroscopy, fluorescence spectroscopy, luminescence spectroscopy, internal reflection spectroscopy and surface plasmon spectroscopy [15]. These converters are very useful for remote analysis because the optical signal is resistant to electrical noise.

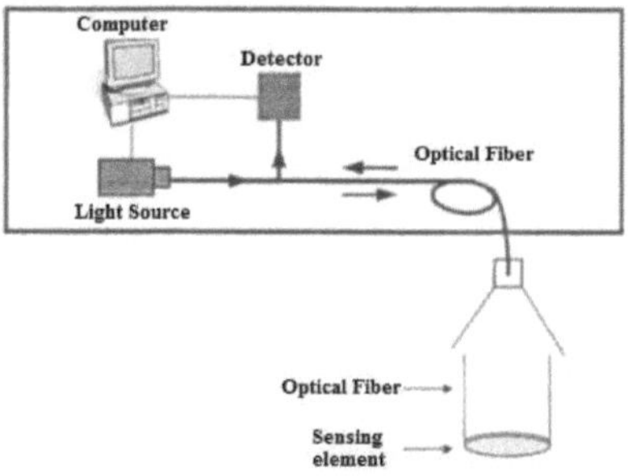

Figure (1): Optical Transducer Based Biosensor [16]

Figure (1) shows an example of an optical transducer based sensor. In this sensor, the sensor part contains substances whose optical properties change with the concentration of the material being measured and the changes made by the optical fiber are transmitted to the detector.

Piezoelectric tools

Piezoelectric tools are based on the current produced by vibration in a crystal or membrane. The vibration frequency is affected by the mass absorbed on the surface. This effect can be linked to an active biochemical reaction [11]. The surface acoustic wave method uses a piezoelectric crystal such as lithium niobate (LiNbO3), but the waves are produced not in the thickness of the crystal but in its surface [11].

Thermal detection methods

These biosensors are known as enzymatic thermistors, because the heat of enzyme reaction is measured as a function of temperature changes against the concentration of the analyte. Thermal methods are based on the production or absorption of heat in chemical and biochemical processes. This reaction heat is due to the interactions between the enzyme and the analyte. This heat can be measured with a sensitive thermistor and attributed to the reaction rate [11].

2-5 Introducing biosensor generations

Since three different mechanisms of electron transfer are seen in amperometric biosensors, three different generations of these biosensors have been introduced.

2-5-1 First generation

Redox enzymes (oxidizing or reducing substrates) are commonly coupled to amperometric converters. The first generation of amperometric biosensors is based on the measurement of the enzymatic reaction product (hydrogen peroxide, NADH), or on the cofactor (oxygen) consumption.

2-5-2 Second generation

In the second generation of amperometric biosensors, we employ an electron transfer mediator between the redox enzyme center and the electrode surface. Intermediates may be organic or inorganic compounds from transition metals, organic matter, and conductive polymers. Intermediates are relatively small molecules where the redox center of an enzyme or a product of an enzyme reaction (for example with H2O2, NADH, etc.) can be reacted, oxidized or regenerated. Oxidation / reduction form of intermediate molecules at low potentials, oxide / reduction at the electrode surface.

2-5-3 Third generation

Here, the electron transfer between the active center of the enzyme and the electrode surface occurs directly. Only a few enzymes are known to induce such behavior, as most of their active sites are surrounded by a protein peel. This shell is a barrier against electron transfer. HRP can transfer electrons directly because the redox center is located near the protein surface.

2-6 Parameters affecting the performance of sensors and biosensors

2-6-1 Response scope and linear scope

For any measurement method, it is important to know the concentration range of the material to which the measurement method responds and in terms of response ratings, it is important to know what range of response concentration is linear. The lower limit of this linear range is called the detection limit. This issue is illustrated in Figure 2. According to the IUPAC definition, the detection limit is the concentration of the analyte that intersects the baseline along the extrapolated linear part of the curve. The baseline for a line is the horizontal line, where the magnitude of the response changes to zero despite the concentration change [6].

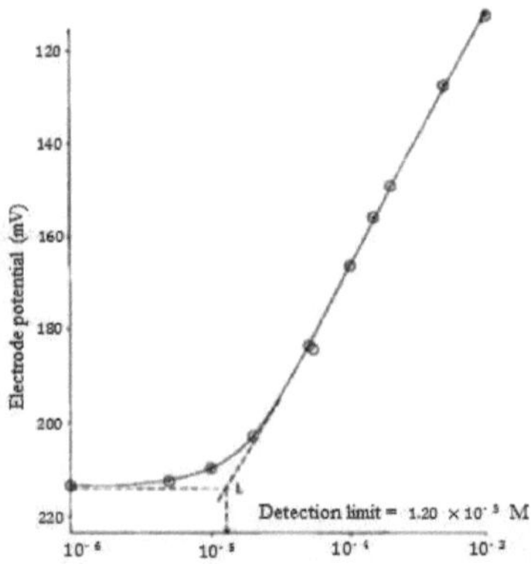

Figure (2): Nitrate Detection Limit [14]

2-6-2 Repeatability

In any measuring instrument, reproducibility is an important factor. This is very important for biosensors, as it is impossible to refill the same quality biomaterials with conventional chemicals.

2-6-3 Response time

Many analytical tools require a settling-down time, meaning they need time to reach equilibrium. This issue is true of many chemical methods. Recent studies, using a nitrate electrode, show that the most repeatable results are obtained after 30 seconds of stirring the test solution in contact with the electrode. In the case of biosensors, the measurement time may be long. Obviously, if this time is prolonged, it will affect the efficiency of the repeated measurements. For biosensors, the response time varies from a few seconds to a few minutes [6].

Recovery time means the time a biosensor has to spend to prepare for the next measurement, which depends on the response time. The recovery time is that after each measurement, the machine must rest until it returns to its original equilibrium before measuring the next sample. In many articles these times are considered together and the total time actually spent on the measurement or the number of samples being measured per hour is reported [11].

2-6-4 Lifetime

All organic material changes over lifetime, especially when they go out of their natural environment. This means that one of the disadvantages of biosensors is that biological components are usually short lived. The construction of all new biosensors involves studies to show how biosensors' response to a standard sample changes over time and over hours, days, and months. Generally, pure enzymes have the least stability, whereas the tissues produced are longer. Three aspects can be studied: the working life of the biosensor, the life of the stored biosensor, and the life of the biomass stored separately [11].

2-6-5 Buffer pH

Precise control of the pH of the test solution is usually necessary. Often a neutral pH phosphate buffer is used. However, pH is highly dependent on the intermediate used for electron transfer.

2-7 Enzyme fixation methods

Chemical methods of enzyme fixation have a longer lifespan. But their response is limited because the process of mass transfer is not easy. Of course, physical methods are the opposite. However, sometimes the chemical method damages the enzyme and causes a greater deficiency in the response, which is of no importance compared to the rapid loss of the enzyme due to poor binding of the physical methods [6].

In this section, different methods of consolidation are presented. The choice of fixation method depends on the type of enzyme and each method has its advantages and disadvantages. During stabilization due to the restructuring of the enzyme, its activity decreases depending on the method of stabilization. The activity of the stabilized enzyme is reduced due to the penetration resistance that reduces or prevents the amount of substrate reaching the active site of the enzyme. The enzymes are applied either in liquid form or by fixation on various bases. In biosensors, the stabilized expression implies an inactive state in which the activity of the enzyme is maintained [17]. The major advantages of the stabilized enzyme include reduced operating costs, greater stability of the enzyme in the stabilized state, the ability to rapidly stop the reaction by simply transferring the enzyme from the reaction solution, repeated use of the stabilized enzyme, and non-contamination of the sample with the enzyme. One of the most important disadvantages of the stabilized enzyme compared to the free enzyme is its low activity and limited mass transfer [18]. The use of biosensors based on stabilized enzymes for analysis purposes is increasing. Increasing the half-life of enzymatic systems and the possibility of eliminating sample preparation has made enzymatic sensors a useful tool in the food and medical industry [19, 20]. One of the important components in enzyme fixation is the availability of a base or support. This base can vary depending on the stabilization process. The basic methods for enzyme fixation in biosensors are as follows:

1. Adsorption
2. Malice
3. Enclosure
4. Cross-linking covalent bond

2-8 Role of nanotechnology in the development of chemical biosensor

In the past decade, much attention has been devoted to the application of nanomaterials such as nanowires, nanotubes and nanocrystals to sensors. This can be due to the ability to organize the size and structure of the sensors as well as the unique properties of the nanomaterials. Thus, great prospects for the design of new analytical systems and increasing the performance of bioassay experiments are opening up [21, 22]. Given that most biological systems such as viruses, membranes, and protein complexes are naturally nanostructured and occur at nanometer-scale molecular interactions, nanomaterials are a suitable candidate for application. They are bio-analytical and medical tools [23, 24]. In addition, nanomaterials pave the way for miniaturizing nanoscale sensors and devices (nanobiosensor and nanosensor) to achieve better sensitivity, specificity, and higher detection speeds than current systems [25].

The chemical, electronic, and optical properties of nanomaterials usually depend on their size and morphology [26]. Various structures have been reported for analytical applications. In the meantime, organic and inorganic nanoparticles and nanotubes and metal oxide nanowires have provided promise for electrochemical nanoscale biosensors due to their biocompatibility and combination of properties Such as high specific surface area, good electrical properties and chemical stability. Over the past decade this branch of science has been extensively studied and numerous articles have been published in this field. The following is a brief description of advances in the last few years (since 2005) in the application of nanomaterials such as carbon nanotubes[1], nanoparticles and polymer nanostructures in chemical bioassay systems.

2-8-1 Carbon nanotubes

CNTs have a fluorine-like structure that can be single-walled and multi-walled nanotubes. SWNTs are cylindrical graphite plates with diameters of 0.5–5 nm with hemispherical end regions, while MWNTs contain several concentric cylinders of these graphite plates with interlayer distances of 0.4 to 0.3-3 nm. MWNTs typically have a diameter of about 2–100 nm.

SWNT *MWNT*

Figure (3): Schematic of Multi-Walled Carbon Nanotubes (MWNTs) and Single-Walled Carbon Nanotubes (SWNTs) [27]

[1] CNTs

2-8-2 Polymeric nanostructures

Another class of nanostructures is polymer nanostructures that allow easier chemical deformation than CNTs. Conductive polymers are multifunctional materials that can be used as receptors, transducers or stabilization matrices in electrochemical bioassays. These conductor polymers are described by the π bond developed during the main building [28]. The electrical conductivity of these polymers results from the formation of charge-doping carriers based on their redox (p-doping) and reduction (n-doping) their main structure [29]. In this way, they mimic the electrical properties of metals, while maintaining the properties of organic polymers such as light weight, corrosion resistance, flexibility and ease of fabrication [30]. As conductive polymers become nanostructured, they find more interesting properties: ease of preparation by chemical and electrochemical methods, sensitivity to a wider range of analytes, and greater electron transfer rate. In addition, it is easy to chemically functionalize their structure to achieve specific selectivity for different compounds, and these are also capable of accepting methods of fabrication that greatly facilitate the miniaturization and production of the array [31].

2-8-3 Nanoparticles

One class of nanomaterials is nanoparticles. Depending on their composition (metal, semiconductor, magnetic), nanosized particles (or aggregates) play different roles in electroanalytical applications. Metal nanoparticles provide three major uses: enhancing the electrical contact between the biomolecules and the electrode surface, catalytic effects, and with semiconductor nanoparticles, labeling and signal amplification [32]. They are typically obtainable by the chemical reduction of the corresponding intermediate salts in the presence of a stabilizer (self-assembled monolayers, microemulsions, polymeric matrices), that provides good surface stability and functionalization to modulate load and solubility properties. Among them, the most widely used gold nanoparticles (AuNPs) are due to their unique biocompatibility, structural, electronic and catalytic properties [32, 34]. Some recent research has focused more on the applications of AuNPs to biosensors using various enzymes including glucose oxidase (GOD), DNA and NADH.

Metallic particles mainly act as essential constituents in the stabilization of biomolecules, separation and transfer of reactants. The magnetic core-shell is usually composed of iron oxides derived from the co-deposition of aqueous salts of Fe (II) and Fe (III) by the addition of a base. By modifying the synthesis parameters, it is possible to achieve the characteristics required for biomedical and bioengineering applications such as uniform size (less than 100 nm), specific physical and chemical properties, and high magnetization. In addition, biodegradability and biocompatibility can be achieved by appropriate surface coating [35].

2-8-4 Nanocomposites

In nanocomposites, the amplifier or filler phase has a high surface-to-volume ratio or an extremely high state ratio. Reinforcing material has been made from particles (eg inorganic, metal oxides), plates (eg exfoliated clay, organic clay) or fibers (eg carbon nanotubes (CNTs) or electrospray fibers). The interface between the matrix and the reinforcement phases is significantly better than the composite material which affects the properties of the matrix material, such as dependence on local chemistry, degree of thermostat hardening, polymer chain mobility, polymer chain formulation or degree of polymer chain sorting [36, 37].

In addition, one advantage lies in the fact that a relatively small amount of filler or amplifier in the nanoscale phase can have a significant effect on the composition of the macroscale. For example, adding CNT as a regenerator can easily improve the electrical and thermal conductivity of the composite material [37, 38].

Other types of nanofillers can improve other properties, such as optical, dielectric, heat resistance, bending resistance, strength, abrasion resistance and damage. In general, nano-amplifiers are scattered in each dimension in polymer, ceramic, or metal matrices during the process.

2-9 Nanostructured coatings

Self-preparation

In nature, for millions of years, methods have been put in place to assemble inorganic nanocrystals at ambient pressure and temperature. This process is called biological mineralization, in which calcium phosphate or carbonate is formed in the bones. A team of researchers attempted to model this process, resulting in a self-preparation method. Self-preparation in chemistry and materials is defined as the simultaneous and reversible organization of molecular units into regular structures by noncovalent interactions. An important feature is the synchronization of the interaction and formation of the organized state, which is widely used in the creation of nanostructured layers on the surface method [39].

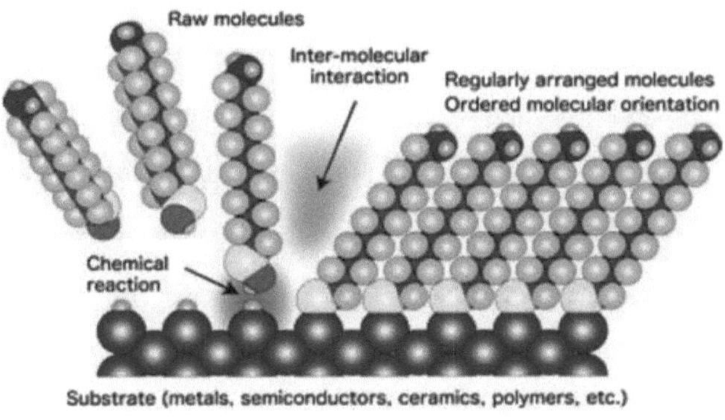

Figure (4): Self- preparation of ordered molecules [39]

Electro-deposition

The process of electro-deposition, also called electroplating, is limited to conductive materials. There are two general methods for layering: 1) Electroplating and 2) Electroless Plating.

In the first method, the material is placed in a liquid solution (electrolyte). When the electrical potential difference is established between the sample electrode and the control electrode (usually platinum), an oxide-reduction reaction occurs, which forms a layer of material on the sample and creates a gas on the control electrode. The electrolyzing process is used to make metal films such as silver, gold, copper, nickel, etc [40, 41].

In the second method, a complex chemical solution is required. In this method, simultaneous deposition is formed on a surface that has an appropriate electrochemical potential with the solution. This method is suitable because it does not require external electrical potential. Unfortunately, in this method it is difficult to control the film thickness and its shape [42].

2-10 Bio-electrochemical enzymatic electrochemical biosensors

The rate of electron transfer between two points depends on the distance between the two points. In bio-electrochemical processes if the electrode surface and the prosthetic group present in the redox enzymes are considered as two electron acceptor and donor centers, the amino acid layers (which surround the redox center of the enzyme) create an energy-efficient barrier to electron transfer. Therefore, the cleavage of the redox enzyme by the amino acid layer prevents direct electron exchange between the enzyme and the electrode. However, under specific and controlled conditions, some enzymes can communicate directly with the electrode without direct contact and in this case, the catalytic reactions of these enzymes are detected by the electrode.

2-11 A review of the research background

This section can be divided into two categories of studies: (1) biosensors and sensors for the detection of various gases; and (2) biosensors and sensors for the detection of nitric oxide (NO).

2-11-1 Biosensors and gas sensors

Electrochemical biosensors are widely used for the analysis of various chemicals (such as glucose, cholesterol, lactate, alcohols, phenols, etc.) in aqueous solution form for environmental, clinical and biotechnological applications. [43]. The great advantage of electrochemical biosensors is based on the biological recognition of proteins, such as enzymes, their specificity and high intrinsic selectivity in the respective substrate analysis.

Recently, some advances have been reported in the field of electrochemical gas biosensors for the detection and monitoring of gas or air gas substrates [43]. Electrochemical gas biosensors based on the micro-electrochemical system consist of two sets of gold microbands for the detection of SO2, phenol, ethanol and formaldehyde. A gel layer containing the stabilized enzyme and electrolyte were applied to the working area of the microelectrode system. Enzymes such as sulfite oxidase, alcohol oxidase, alcohol dihydrogenase or formaldehyde dihydrogenase were used. Denison et al. [44] designed a similar biosensor to measure the concentration of phenol vapor using polyphenol oxidase (E.C 1.14.18.1). This biosensor has a very high sensitivity, but its normal operation is largely influenced by the relative humidity of the air. Denison and his colleagues observed a decrease in the biosensor flowmeter signal in all cases, as the air humidity decreased, resulting from the rapid loss of water from the gel containing the enzyme.

Another type of gas biosensor for ethanol was reported by Mitsubayashi et al. (1994) [45], containing a two-part cell (for liquid and vapor phase) separated by a porous PTFE diaphragm membrane. The enzyme electrode was fabricated using a commercial oxygen-soluble Clark-type electrode in which the oxidase enzyme was stabilized. The gas sensor showed high selectivity for ethanol. The observed short life of this gas sensor is due to the inactivation of alcohol oxidase and its loss from the immobilized membrane.

A gas biosensor was proposed by Hämmerle et al. (1996) [46] for the detection of formaldehyde in the atmosphere. This cell contained the FDH enzyme (formaldehyde dihydrogenase (E.C 1.2.1.46)) that acted directly in the gas phase with an almost linear response to formaldehyde up to a concentration of 6 vpm. The detection range was 0-3 vppm. In addition to formaldehyde at concentrations of 2 vppm and 6 vppm, TB was able to function for at least 7 hours without loss of activity. This electrochemical cell comprises a 3-electrode system in a plastic chamber (Fig. 5). Black platinum or graphite attached to a porous Teflon membrane (Neotronics, 2.6 cm in diameter) was used as the working electrode and graphite attached to the same membrane as the auxiliary electrode. The geometric area of the two-dimensional electrode was 2 cm2, regardless of the edges of the membrane. The electrodes are in contact with the platinum strips. The reference electrode was a saturated white mercury electrode (+ 244 mV vs. NHE +) and a gas inlet diameter of 1.5 cm.

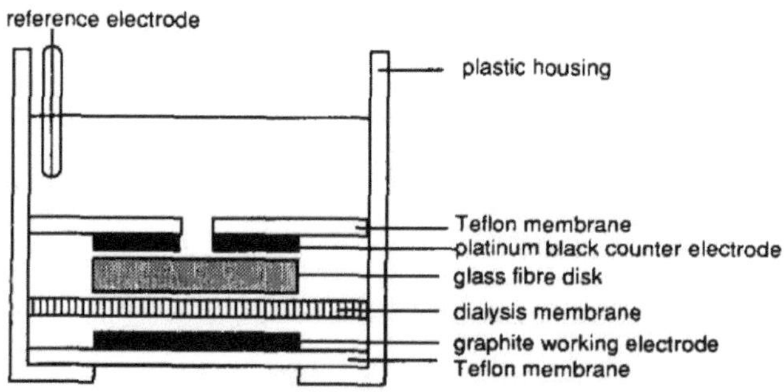

Figure (5): Schematic drawing of electrochemical cell. Working and auxiliary electrodes are in contact with a platinum strip (not shown) [46]

This design ensures high loading of the enzyme in the vicinity of the working electrode, whereby the analyte enters the electrochemical cell.

One of the most recent studies on formaldehyde identification has been reported by Fauzia et al. [47] (2018). They presented a gold-based biosensor with gold nanoparticles to detect formaldehyde gas. The gold nanoparticles were synthesized on the surface of a glass substrate and then coated with a polyacrylate membrane film containing Chromioniphor I (5294ETH) and alcohol oxidase enzyme. The basis of this biosensor is based on chromo-color change. This biosensor performed well in formaldehyde detection, with a wide linear response range of M.1–7-10 and a low detection limit of M.1.

2-11-2 Biosensors and sensors NO detection

He et al. (2009) [79] used a nano-alumina modified glass carbon electrode (Al2O3) to detect NO. Cyclic voltammetry, chronoamperometry and linear scanning voltammetry were used to investigate the electrochemical behavior of the sensor made in reaction with NO. SEM was used to study the surface morphology of the electrode, which showed that the alumina film was well formed on the electrode surface. Under optimal conditions, the NO oxidation peak current increased linearly in the concentration range of 0.04–0 210 M and reached a detection limit of 7.2 nM. In addition, this sensor showed high stability, reproducibility, and selectivity against NO detection [48].

Deng et al. (2010) [49] have been presented an electrochemical sensor with high sensitivity and good selectivity for NO detection. The sensor was fabricated by placing a multi-walled carbon nanotube / chitosan / gold nanotube layer (Na fi on / MWNTs-CS-AuNPs) on the glass carbon electrode surface. Deng et al have been investigated several parameters affecting the electrochemical response such as film thickness, applied potential and Nafion volume and reported their optimal amount. The sensor showed a linear range of 0/019/54 with a detection limit of 0.0076.

Bai et al. (2017) [50] have been presented a graphene-based nanobiosensor in which they used a sonochemical method to synthesize graphene-gold (G-Au) nanocomposites. During the sonochemical procedure, simultaneous grinding of graphite and gold chloride reversal occur to produce highly crystalline G-Au nanocomposites. In this process, the growth of gold nanoparticles occurs on the surface of laminated graphene sheets. These nanocomposites were characterized using Uv-vis, XRD, FTIR, TEM, XPS and Raman spectroscopy. For example, the results of the UV-vis analysis of the synthesized nanocomposites are shown in Figure (6).

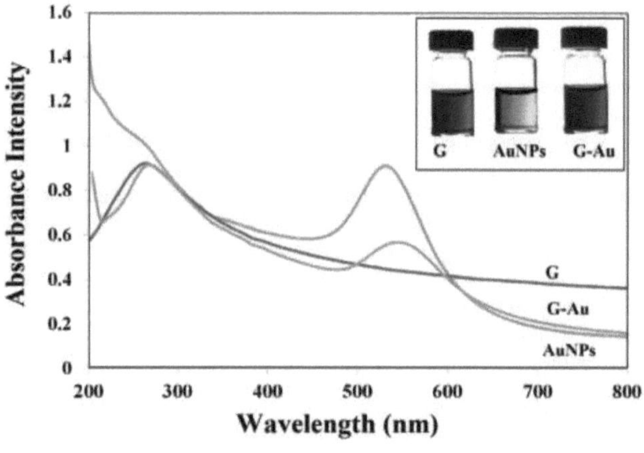

Figure (6): UV-vis spectroscopy analysis of graphene (G), gold nanoparticles (AuNPs) and graphene-gold nanocomposite (G-Au) [50].

In the case of G-Au, the LSPR peak was slightly redshifted, indicating the deposition of gold nanoparticles on graphene plates. Bai and his colleagues investigated the performance of their sensors using cyclic voltammetry. As can be seen in Figure (7), the electrode with G-Au nanocomposite exhibited the largest oxidation peak, indicating that the produced nanocomposite causes increased NO oxidation rate.

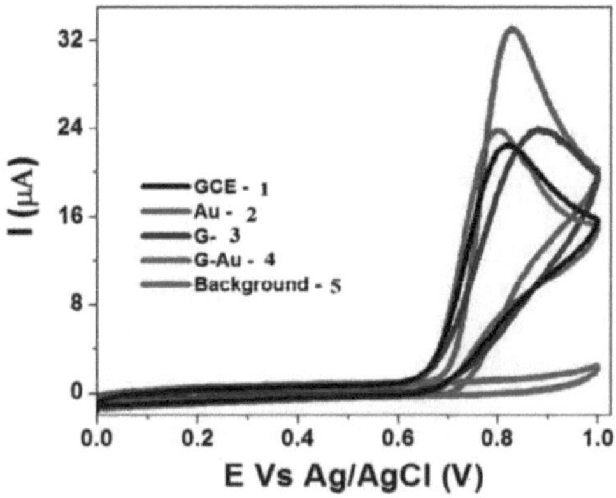

Figure (7): Cyclic graphs of G-Au / GCE electrode in buffer solution (2.5 pH) of 0.1 M PBS at 1 mM NO concentration at 100 mV / s scanning rate [50].

The calibration graph reported by Bai et al. Comprises three linear response ranges (Figure 8). It is observed that in the range of lower concentrations (1-10) the sensitivity of the sensor (0.1212) is much higher.

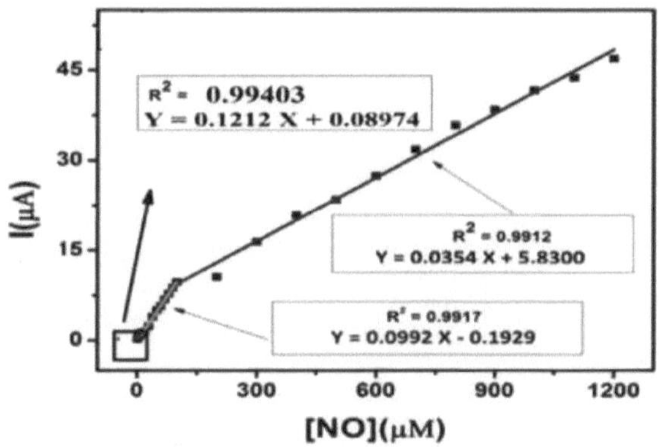

Figure (8): Calibration diagram of the G-Au / GCE electrode to detect NO [50].

CHAPTER

3

Research Method

3-1 Introduction

In this study, we have been investigated the performance of two cytochrome c oxidase (Cyt c oxidase) and cytochrome c protein (Cyt c) enzymes and a graphene-based or reduced graphene oxide (rGO) nanosensor for detecting nitric oxide (NO) gas.

Nitric oxide (NO) is a colorless gaseous chemical compound. It is a compound with the chemical formula NO of a free radical. In mammals such as humans, NO is an important molecule in cell signaling that is involved in many physiological and pathological processes. This compound has a short half-life in the blood of a few seconds (less than 6 seconds) [51].

In general, NO has been studied due to its effects on air pollution and in particular biological systems. Abnormal production of NO causes some important biological activities and leads to many diseases [52]. Deficiency or overproduction of NO can lead to hypertension, stroke, Parkinson's disease, Alzheimer's disease and diabetes (types I and II). In biochemical and especially medical approaches, the measurement of NO produced in normal and abnormal tissues, including in vivo or in vitro measurements, is essential. The purpose of this study was to fabricate nanobiosensors and nanosensors for rapid and accurate identification of this compound. Therefore, in this study NO is the analyte evaluated and is produced in the laboratory by the conventional method (described below).

In this study, cytochrome c oxidase (EC: 1.9.3.1) was used. This enzyme is a metalloenzyme (containing copper and iron), present in the mitochondrial (or bacterial) respiratory electron transport chain located in the mitochondrial (or bacterial) membrane. They receive the electron from Cyt c and transfer it to an oxygen molecule. Molecular oxygen is then converted to two water molecules. Reaction (1) indicates the enzyme function.

$$4\,Fe^{2+} - cytochrome\;c + 8\,H^{+}_{in} + O_2 \rightarrow 4\,Fe^{3+} - cytochrome\;c + 2\,H_2O + 4\,H^{+}_{out} \quad (1)$$

Cyanide, sulfide, azide, and carbon monoxide all bind to Oxidase Cyt c, thus inhibiting the function of the enzyme, leading to chemical strangulation of the cells. This enzyme was prepared by U 20 activity from Sigma Aldrich. This enzyme was stored at -20 ° C. The stable form of the enzyme, which is soluble (5 mg protein / ml), has been used. Due to its high price, low content, and unsustainable activity, a small amount of this enzyme was dissolved in a small amount of distilled water (100 µl) and kept in the refrigerator, but the new solution was replaced once every 3 days.

Detector chemical element (Graphene)

The small size and the electronic structure of the carbon make it a special element capable of producing multidimensional structures with remarkable properties. Graphene is a two-dimensional (one-atom-thick) allotrop of carbon with a planar structure like a honeycomb, this material being one of the strongest materials ever measured (figure 9) [53]. In this study, graphene was used in the manufacture of graphene-based sensors to detect NO. Chemical and electrochemical synthesis methods have been used to produce it.

Figure (9): A single graphene plate [53]

3-2 Main materials of working electrode modifying nanocomposites

Gold working electrode modifications were used to better detect NO in order to improve electron transfer properties (such as gold nanoparticles (AuNPs)) and to create a stable and compatible environment (such as chitosan (CS) and polyaniline (PANI)) with biological element. Nafion was also used to prevent the intruders from penetrating the identification process. The following is a brief introduction to these materials.

3-2-1 Gold Nanoparticles (AuNPs)

Due to the new properties and properties of nanometer-sized materials in the industry, there is a great desire today for their processing and application. AuNPs are a class of nanoparticles with a long history of use. Colloidal gold nanoparticles are colloidal suspension of AuNPs in a fluid and usually water. The liquid is dark red (for particles less than 100 nm) or blue / purple (for larger particles). Due to its unique optical, electronic, and molecular detection properties, AuNPs have attracted the attention of many researchers and are used in various fields such as electron microscopy, electronics, nanotechnology and materials science. The properties of gold colloidal nanoparticles depend on their size and shape. In this study, citrate was used to synthesize AuNPs.

3-2-2 Chitosan (CS)

Chitosan organic polymer is a linear polysaccharide that is derived from the chelate diacetylation in the outer skeleton of insects, arthropods, crustaceans (such as shrimp, crabs) and the cell walls of fungi. Chitosan Organic Polymer is a linear polysaccharide with excellent membrane forming ability, good biocompatibility, high water permeability, adhesion, non-toxicity and high mechanical strength making these properties one of the best options. It is designed for the stabilization of enzymes or proteins and therefore for use in biodegradation applications [54]. A schematic of the chemical structure of chitosan is presented in Figure 10. Low molecular weight chitosan (85% diacetyl) prepared by Sigma Aldrich USA was used in this study. Chitosan is insoluble in pHs above 6.3 due to de-protonation of amino groups and soluble in acidic gels.

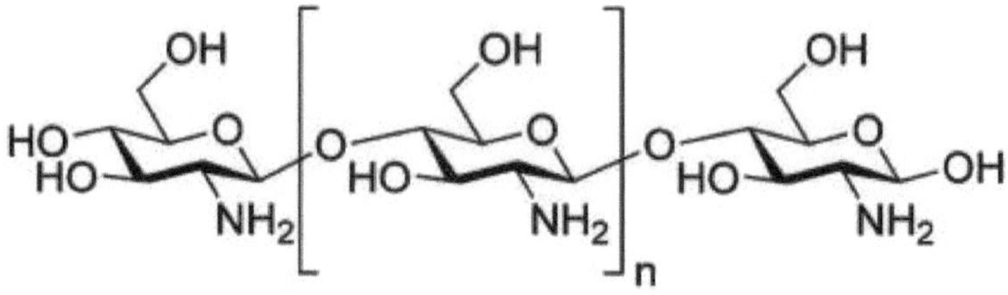

Figure (10): Chitosan structure

3-2-3 Nafion

Nafion is a cation exchange membrane, and anions cannot pass through it. Nephi passes NO. Nafion is used in this study to prevent intrusions such as nitrite (NO2-), hydrogen peroxide (H2O2) and ascorbic acid (AA) as well as to prevent enzyme leakage from the electrode surface. In this study, Sigma Aldrich's Nafion 117 solution (5% by weight) was used to prepare the final Nafion solution.

3-2-4 Polyaniline (PANI)

Among the various conductor polymers, polyaniline is a unique item due to its easy synthesis, high conductivity, stability and reasonable price. The structure of the polyaniline is shown in Figure 11.

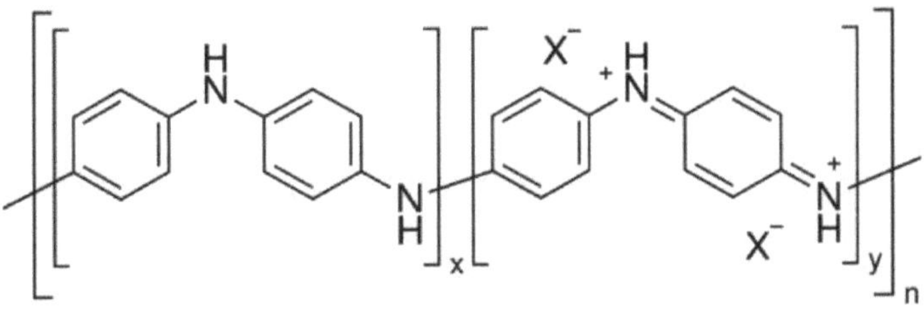

Figure (11): Structure of polyaniline

Although the methods of synthesis of polyaniline are very simple, the mechanism of aniline polymerization is very complex.

3-3 Equipment & Devices

3-3-1 Potentiostat device and three-electrode electrochemical cell

The Potentiostat model SP-150 from BioLogic Science Instruments was used in this work to perform ampometric and voltammetric tests. Experiments were performed on a three-electrode handmade cell.

3-3-2 Electrodes

Reference electrode Ag / AgCl, platinum auxiliary electrode and gold handmade electrode were used to detect NO. Due to the high gold electrodes available in the market and the faster preparation of modified electrode samples for morphological analysis, 6 gold working electrodes were prepared manually.

3-4 Manufacture of standard nitric oxide (NO) gas solution

The standard NO solution is produced by adding 2 molar H2SO4 droplets to a glass container containing saturated NaNO2 solution [55]. To remove oxygen and nitrogen oxides, the gas resulting from this reaction is continuously passed through 5% (w / v) pyrogel solution in saturated potassium hydroxide and 10% potassium hydroxide (w / v), respectively. Pyrogol is an oxygen absorber. To produce the NO saturation solution, the gas passing through these adsorbents was finally bubbled into a 10 mL glass jar containing 0.1 M PBS (7pH) containing 0.1 M NaCl. This process will continue for 30 minutes until it finally forms in the final container of the NO saturation solution at a concentration of 1.8 mM [55]. The side containing the NO-saturated solution should be covered with foil and sealed carefully to prevent oxygen penetration into it. All of these steps are performed in the Globox container. Before the process begins, nitrogen gas must flow through all equipment for 30 minutes to make the system oxygen-free. Because NO reacts quickly with oxygen. Since NO is a toxic compound that is lethal at concentrations above 100 ppm, it is best to perform all the preparation and storage steps under the chemical hood.

Standard NO solutions for injection into the electrochemical cell are produced by the sequential dilution of the primary NO solution during the NO detection process.

3-5 Electrochemical Methods

3-5-1 Electrodeposition

The electrolytic process is a method of precipitating and placing the conductive material on the surfaces. In this method the material is usually placed in a liquid solution (electrolyte). When the electrical potential difference between the working electrode and the reference electrode is established, an oxidation-reduction reaction occurs that precipitates and settles on the sample.

3-5-2 Amprometric method

Amprometric method is a suitable method for the detection of analytes capable of oxidation or reduction in chemical and biochemical reactions. In the amperometric method, a potential signal is generated by applying the potential difference between the reference electrode and the working electrode. Amprometry is performed in a carrier electrolyte solution (including buffer solution at constant pH) to provide electron transfer to the electrical circuit. Flow amperometric biosensors generate proportional to the analyte concentration studied [56]. Amprometry may be performed at a fixed potential or differential pulse potential. Chronoamprometry is a type of ammeter that responds at a constant potential under unstimulated conditions. The reason is to stir the solution to prevent mass transfer by the displacement method and to provide natural mass transfer at the electrode surface.

Voltammetry is an effective technique for studying the electrochemical behavior (redox) of biosensors in the vicinity of the analyte. In voltammetry, the current is measured under conditions where the potential difference between the reference electrode and the working electrode is changed. Linear Scanning Voltammetry (LSV) is a method in which the applied potential is linearly changed and the response current is displayed in terms of potential difference in the voltamogram diagrams. Cyclic voltammetry (CV) is another method that examines the oxidation and reduction in chemical reactions. In CV, the potential difference with a constant scanning speed is changed from the initial value to the secondary value and then returned to the secondary value again.

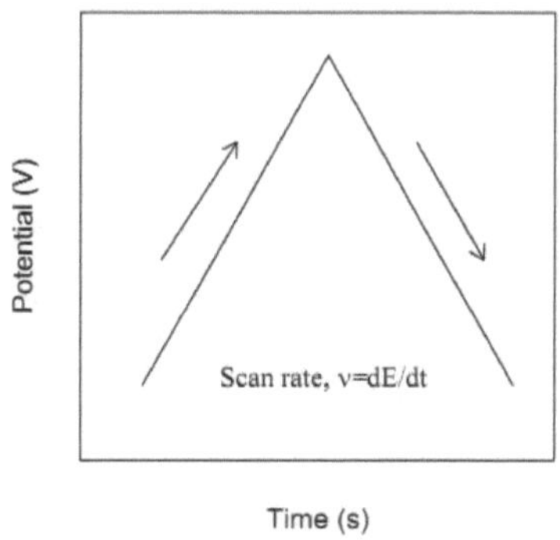

Figure (12): Voltamogram of an oxidation-reduction reaction on the electrode surface in a biosensor [56]

Figure (12) shows the increase and decrease of the potential over time for the CV on the surface of a biosensor. In this diagram, as can be seen, the cyclic voltammetry consists of two linear scanning stages in which the potential scanning speeds are equal and different. Given that the electrolyte analyte at the biosensor surface participates in an oxidation-reduction reaction, the half-oxidation-reduction reaction is given as follows:

$$Ox + ne^{-} \leftrightarrow Red \qquad 2$$

Ox and Red are oxidized and reduced in the oxidation-reduction reaction, respectively. Figure 13 shows the cyclic voltamogram of a reversible oxidation-reduction reaction in the vicinity of an electrochemical biosensor.

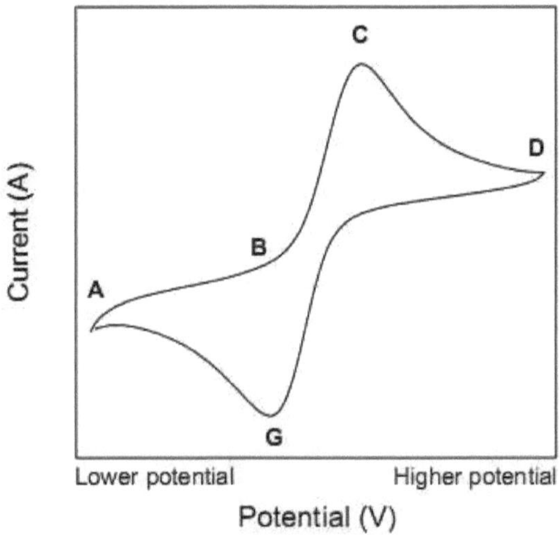

Figure (13): Reversible Oxidation-Reduction Cycle Voltamograms in the Vicinity of the Electrochemical Biosensor [56]

The starting point (A) is low. From point A to B, the flow response increases with a slight slope, that is called background flow. The field flow is related to the residual current of impurities as well as the charge of the bilayer. At point B, the potential for oxidation of the analyte is provided, thus reducing the oxidized species. At this point, the flow response increases with increasing potential value and reaches the peak anodic current shown at point C. After the analyte was oxidized in the vicinity of the working electrode; the current response to point B decreases. For a reversible reaction, reversal of the oxidized species occurs in the reverse scanning phase and a cathodic peak current (point G) is obtained. In cases where the potential is initially scaled by the negative slope, the anodic and cathodic peak shifts.

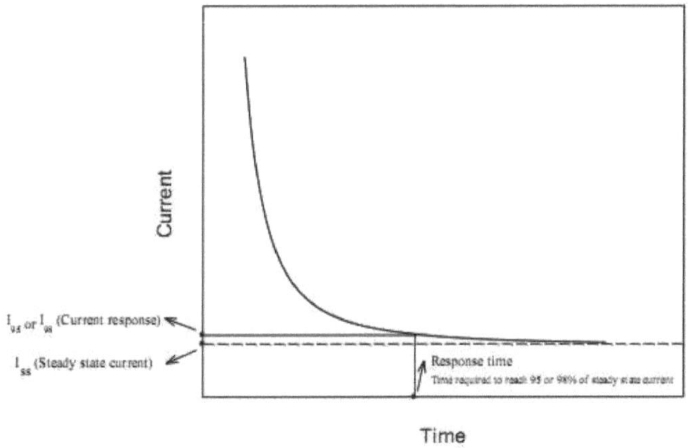

Figure (14): Chronoamprometry: Current Response in Time [56]

The value of the peak difference corresponding to the peak current obtained in the CV method is valuable information used in ampometric detection. If the potential difference between the reference electrode and the working electrode is set to the peak potential, the analyte base current is increased and the intracellular current decreases, thus increasing the sensitivity and accuracy of the electrochemical detection.

Figure (14) shows a typical flow response for the chronoamprometric method. Initially, the flow rate is high and then the flow decreases to reach a steady value. This is due to the fact that the amount of substrate near the electrode is high at the onset of the substrate, following the participation of the substrate in the biochemical reaction and the stability of the substrate concentration gradient at the electrode surface and in the mass fluid (diffusion and expansion of the diffusion layer) and decrease in force. The impulse drive decreases the flow rate and reaches a constant value. In the steady-state region, the current response is time independent.

In practical terms, it is much more accurate and practical to determine the concentration of an analyte using a calibration graph. In this method, an amperometric test is performed at different concentrations of the analyte. Then the steady state response flow is plotted in terms of concentration and calibration graph. Another common method for calibration diagrams is by first performing an amperometric test for the carrier electrolyte solution, allowing the field flow to subside and reaching steady state, and then determining the amount of analyte from solution to solution. The carrier was added to various concentrations of the analyte and the flow rate and current changes were recorded continuously. In this method, stirring conditions are needed to facilitate the mass transfer to the solution. Figure (15) shows the amprometric response for injection of different concentrations of analyte over time.

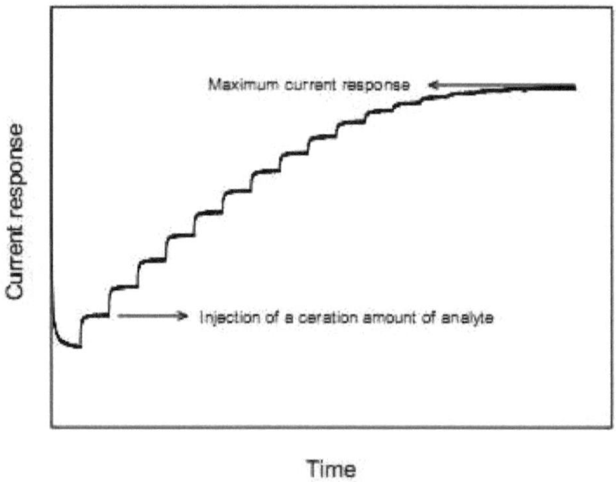

Figure (15): Amperometric flow response for increasing concentrations of analyte in solution over time [56]

Figure (16) presents an amperometric flow response graph in terms of analyte concentration for the enzyme biosensor. In enzymatic biosensors, the amount of enzyme stabilized at the electrode surface reaches a maximum value, depending on the amount of enzyme stabilized. The linear area of the response profile of the current is used as a linear calibration graph for analytical measurements.

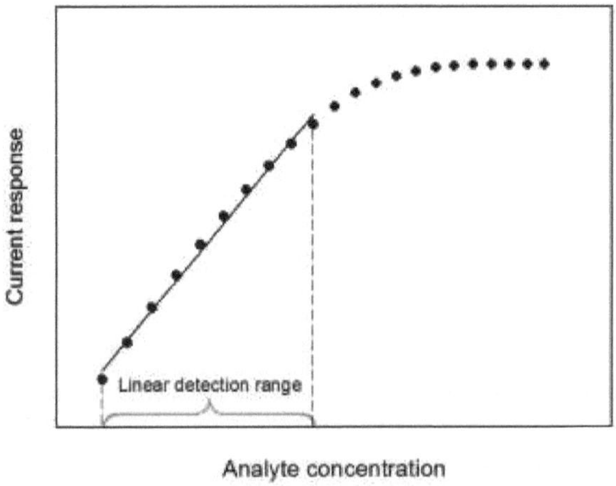

Figure (16): Amperometric flow response in terms of analyte concentration (Calibration Chart) [56]

3-5-3 Electrochemical impedance spectroscopy method

The recording of electrode surface resistance changes during the process of correction by electrochemical impedance spectroscopy (EIS) technique can provide good information on the electrical conductivity of the stabilized layers on the electrode surface. Most of the data obtained by the EIS is graphically expressed in a bad diagram or a Nyquist chart. An example of diagrams derived from the EIS method can be seen in Figure (17). What is important in analyzing this graph is the minimum point in the graph, expressed as the load transfer resistance (Rct). Whatever, this resistance is higher; the conductivity of the investigated surface is lower.

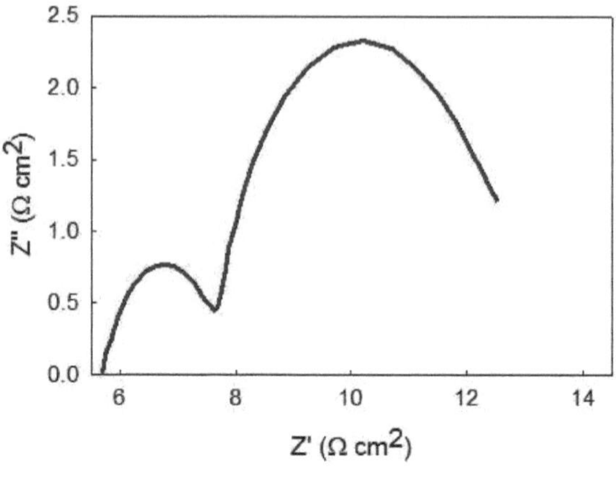

Figure (17): An example of an EIS diagram for surface guidance analysis

3-6 Methods of characterization and electrochemical characterization tests and analyzes

3-6-1 Preparation of UV-vis absorption spectroscopy analysis samples

UV-vis absorption spectroscopy is commonly used to evaluate consistent protein changes and interactions between proteins and other compounds. MPA-AuNPs-Cyt c nanocomposites were prepared by mixing modified Cyt c solution and MPA-AuNPs solution for 1 h. In one such method, the CS-MPA-AuNPs-Cyt c nanocomposite was prepared by mixing the modified Cyt c solution and the CS-MPA-AuNPs solution.

The samples were prepared for UV-vis analysis: quartz quartz was incubated for 30 min in pyrana solution. It was then washed with distilled water and dried at room temperature. Finally, similar amounts of AuNPs, MPa-AuNPs-Cyt c, and CS-MPa-AuNPs-Cyt c samples should be shaken in 3 similar quartz quarts and stored at 4 ° C for 6 h and then stored in a UV-light apparatus. But according to the quartz cuvette, samples cannot be stored in the refrigerator and tested simultaneously. Therefore, each sample was stored separately in the refrigerator and after testing the next sample was poured into the cuvette.

In this study, cyclic voltammetry (CV) technique was used for two purposes:

- Electrolytic
- Investigation of electrochemical behavior of modified electrodes

For each of these cases, in addition to the modified working electrode in question, reference and auxiliary electrodes are also included in the solution. The potential difference between the working electrode and the reference electrode, the scanning rate and the number of cycles are applied through the Potentiostat software. Information on the magnitude of the potential difference applied, the scanning rate and the number of cycles are presented in Chapter 4 before the results are interpreted.

CHAPTER

4

Results and discussion

4-1 Introduction

This chapter discusses the results of experiments. The performance of each sensor is evaluated separately. In general, the evaluation of the morphology and electrochemical performance of the built-in sensors is discussed.

As can be seen in Figure (18), AuNPs have a peak characteristic surface plasmon resonance (LSPR) at 525 nm. However, the absorption spectrum of MPA-AuNPs-Cyt c showed an LSPR peak at 529 nm and a weak Soret peak for protein at 410 nm. The Soret peak is mainly due to the dipole of the electrons. As previously described, the presence of MPA on gold nanoparticles results in its strong binding to Cyt c and changes the dielectric constant of its surrounding environment. These changes in dielectric constant resulted in 4 nm redshift at the LSPR peak for MPA-AuNPs-Cyt c.

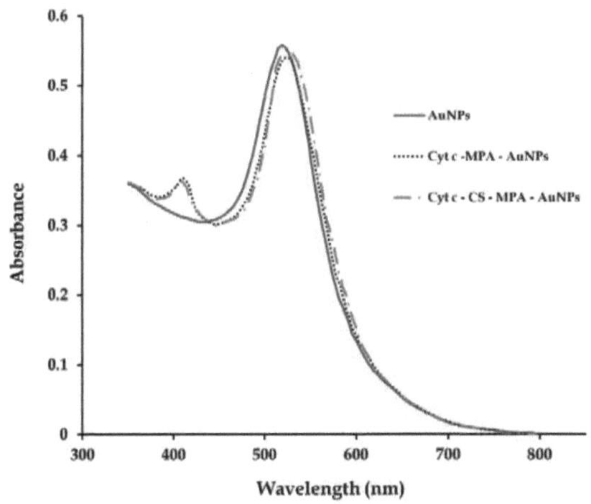

Figure (18): UV-vis absorption spectra for AuNPs, Cyt c-MPa-AuNPs and Cyt c-CS-MPa-AuNPs

The CS-MPA-AuNPs-Cyt c sample showed two peaks at 408 and 532 nm. In addition, the presence of CS resulted in a redshift (ca. 3 nm) in the LSPR peak relative to the MPa-AuNPs-Cyt c sample, which was due to the strong binding of CS to the carboxylic group of MPa-AuNPs. Due to the specific properties of CS mentioned earlier, this material can strongly absorb AuNPs. In addition, CS by its MPA bound inhibits AuNPs density, which results in an increase in LSPR peak (about 5%) of CS-MPA-AuNPs-Cyt c sample compared to MPa-AuNPs-Cyt c sample. Be. However, there was no significant change in the Soret peak for the CS-MPa-AuNPs-Cyt c sample compared to the MPa-AuNPs-Cyt c sample. This indicates that Cyt c can retain its biological catalytic activity in CS-MPA-AuNPs nanocomposites.

4-2 Effect of Cyt c-based electrode modification on voltammetric response and electrochemical impedance spectroscopy

To investigate the effect of electrode modification on biosensor performance, cyclic voltammetry (CV) was used. Figure (4-2) shows the diagrams of the electrodes modified with different layers in 0.1 M PBS buffer (7pH) at 100 mV / s scanning rate. As shown in the figure, no redox peak was observed for the raw gold electrode (curve a). The CV diagram of the Cyt c / MPA / SAMs-Au electrode (curve b) showed a redox peak pair with anodic peak potential of +0.02 02 and -0.045 cathodic peak potential. The peak separation potential of this electrode was 0.064. This indicates that Cyt c is a desirable protein for electron transfer. After adsorption of AuNPs and CS on Cyt c / MPa / SAMs-Au electrode, Cyt c / MPa-AuNPs / SAMs-Au electrodes (curve c) and Cyt c / CS-MPa-AuNPs / SAMs-Au (d curve, potential Anodic: V 006 +, cathodic potential: -0.0443 and peak separation potential V037) showed strong redox peak pairs that increased peak current compared to Cyt c / MPA / SAMs-Au electrode. These results showed that electron transfer was improved by CS-AuNPs nanocomposite. It was observed that the Cyt c / MPa / SAMs-Au (b curve) and Cyt c / MPa-AuNPs / SAMs-Au (c curve) electrodes had a smaller redox peak than the Cyt c / CS-MPa-

AuNPs / SAMs-Au electrode (curve). Curve d) showed. This may be related to the coefficient of transmission between the receiver and the electron transmitter, which strongly depends on the distance between the electron transfer systems [57]. Therefore, it can be said that the structure of CS-AuNPs nanocomposite can significantly reduce the distance between the Cyt c redox centers and the gold electrode. This leads to a significant increase in the effective surface area of the modified electrode to stabilize the greater amount of Cyt c and also to provide high conductivity to improve the electron transfer between the Cyt c and the gold electrode surface. Also, CS-MPA-AuNPs nanocomposite creates a biocompatible microenvironment to maintain Cyt c activity.

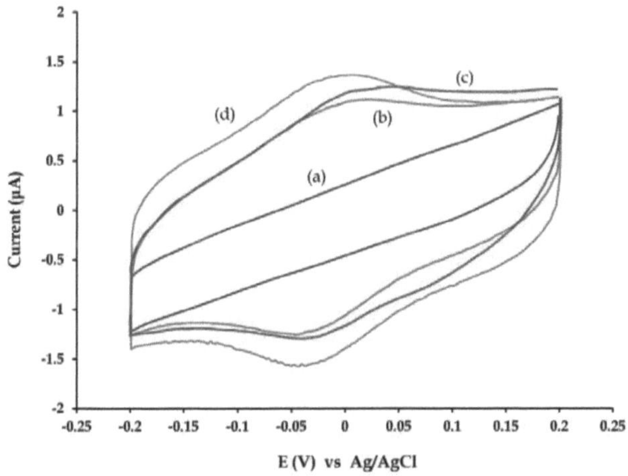

Figure (19): Voltage diagrams: (a) raw gold electrode, (b) Cyt c / MPA / SAMs-Au electrode, (c) Cyt c / MPA-AuNPs / SAMs-Au electrode, and (d) Cyt electrode c / CS-MPA-AuNPs / SAMs-Au in 0.1 M PBS (7pH) at 100 mV / s scanning rate

As mentioned, changes in the resistance of the electrode surface due to its modification with the composite layers can provide useful information about these layers. The electrochemical impedance spectroscopy (EIS) technique can be used for this study. The electrodes modified by this technique were tested in a buffer solution of 5 mM saturated potassium ferricyanide nitrogen ([Fe (CN) 6] 3.4−) containing 0.1 M KCl and the results are shown in Figure 20.

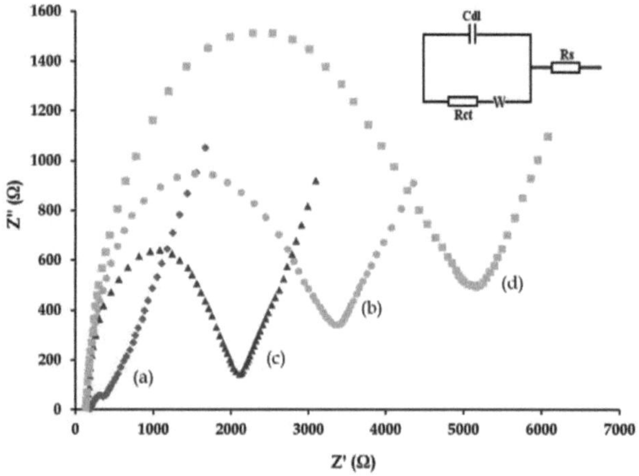

Figure (20): Impedance diagrams: (a) raw gold electrode, (b) SAMs-Au electrode, (c) CS-MPA-AuNPs / SAMs-Au electrode, and (d) Cyt c / CS-MPa- AuNPs / SAMs-Au in [Fe (CN) 6] 3− / 4− 0.005 mM containing 0.1 M KCl at 100 mV / s scanning rate

The charge transfer resistance (Rct) of the SAMs-Au electrode (curve b, Ω 3380) was significantly increased compared to the raw gold electrode (curve a, Ω 355). After the formation of the CS-MPA-AuNPs nanocomposite layer on the SAMs-Au electrode, the strength of the modified CS-MPA-AuNPs / SAMs-Au electrode decreased (c curve, Ω 2110). This indicates high conductivity of CS-MPA-AuNPs nanocomposites. After the addition of Cyt c to this electrode, the final modified electrode experienced a sharp increase in charge transfer resistance up to Ω 5160 (d curve). These results showed that CS-MPA-AuNPs nanocomposite significantly improves electron transport due to its high conductivity.

4-3 Investigation of electrochemical behavior of Cyt c-based biosensor against NO reduction

The CV test was performed in the potential range of 1 to V 0.2 (scanning rate 100 mV / s) in the absence and presence of different NO concentrations. The diagrams of the Cyt c / CS-MPA-AuNPs / SAMs-Au electrode diagrams are visible in the presence of NO in the 0.1 M (7pH) PBS buffered solution in oxygen (Figure 21). According to Figure (21), the redox potential of Cyt c on the electrode surface was about 0 (zero) volts. However, according to the reported studies, the NO reduction peak in the presence of Cyt c is in the range of -0.7 to - 0.8 V [58].

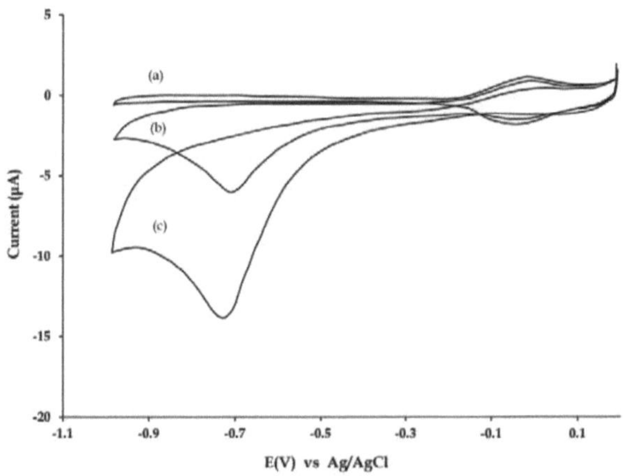

Figure (21): Nafion / Cyt c / CS-MPA-AuNPs / SAMs-Au electrode cyclic diagrams in 0.1 M PBS buffer at different concentrations of NO: (a) 0 µM, (b) 12.5 µM, (c) 50 µM at 100 mV / s scanning rate

As is clear from Figure 21, the reaction between Cyt c and NO shifts the resuscitation peak from -0 to -0.7 V. It is observed that as the concentration of NO increases, the size of the resuscitation peak also increases, indicating the catalytic process of NO reduction (b and c curves) relative to the absence of NO (curve a). These results confirm that Cyt c plays a very catalytic role in NO recovery. The results showed that the built-in biosensor is capable of detecting and quantifying NO over a wide range of concentrations. The role of the CS-MPA-AuNPs composite layer in reducing the potential difference of resuscitation in addition to helping to stabilize Cyt c fixation on the gold electrode surface should not be overlooked. Based on the results, amperometric tests should be performed at a potential difference of -0.7 V relative to the reference electrode.

4-4 Evaluation of the effect of pH on the performance of protein-based biosensor Cyt c

To investigate the effect of pH on the biosensor current response made in the detection of NO50 micromolar, an amperometric test was performed at a potential of -0.7V in the pH range of 5 to 8. As shown in Figure 22, the maximum flow response was obtained at pH 7, which is in fact the optimum pH 7 for Cyt c in solution. This moderate pH provides favorable conditions for electron transfer and Cyt c stabilization.

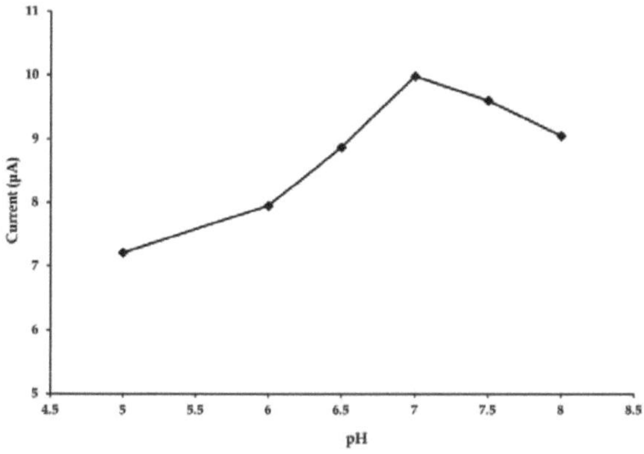

Figure (22): Influence of pH of PBS buffer on Nafion / Cyt c / CS-MPA-AuNPs / SAMs-Au Amperometric Response in Detection of NO 50 µM -0.7 V Potential and Scanning Rate of 100 mV / s

4-5 Amprometric response of Cyt c-based biosensor to NO

After determining the optimum operating conditions, a calibration diagram must be obtained for the biosensor to be applied. To do this, an amperometric test must first be performed under optimum operating conditions. Figure (23) shows the chronoamperometric graph of the Nafion / Cyt c / MPA-AuNPs-CS / SAMs-Au electrode versus NO reduction. The amperometric response of the biosensor was recorded by injecting different concentrations of NO in 0.1 M PBS buffer (7pH) free of oxygen under stirring under gentle conditions. The flow induced by NO reduction was dramatically increased to a constant maximum flow rate and reached within 95% of this maximum within 5 seconds. So the response time of this biosensor is 5 seconds. Such a short response time is due to the high electrical conductivity of CS-MPA-AuNPs nanocomposites. In addition, this nanocomposite is a good effective surface for Cyt c fixation and reduces the distance between the bioactive center of this protein and the electrode surface.

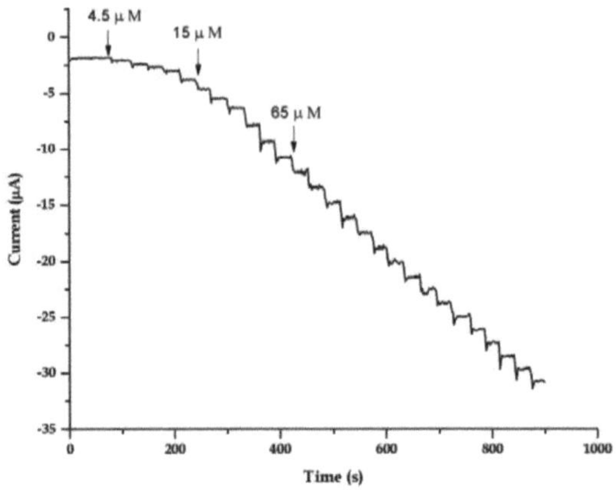

Figure (23): Amperometric Response of Nafion / Cyt c / CS-MPA-AuNPs / SAMs-Au Electrode to Sequential Injections of NO in 10 ml of 0.1 M PBS Buffer at 0.7 V (Relative to the reference electrode)

Using the provided biosensor amperometric response, its calibration curve was plotted (Figure (4-7)). As shown in Figure (24), the linear NO detection range is from 10 to 21μμ (R = 0.999, n = 20). Based on the slope of the calibration line, its sensitivity was 1995 μA / μM (3S / N =). To obtain the detection limit, 10 different biosensor samples were made and then the flow response of each was determined to a specified amount of NO.

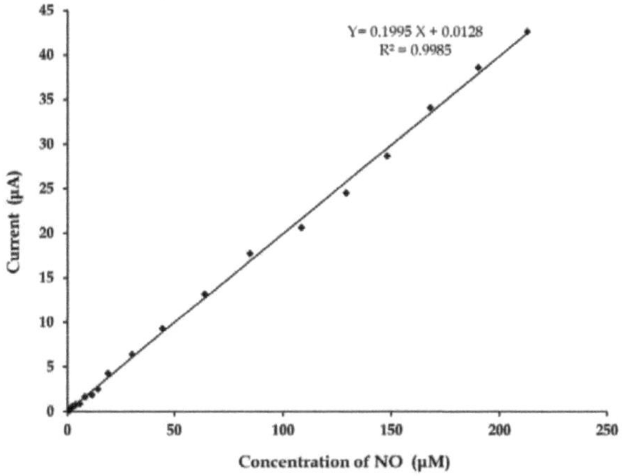

Figure (24): Calibration diagram of Nafion / Cyt c / CS-MPA-AuNPs / SAMs-Au electrode in 0.1 M PBS buffer at -0.7 V potential (relative to reference electrode)

The electrochemical form of the Equinox-Burke equation and the inverse current drawing in terms of concentration were obtained by Michaelis Menten constant (KM) 0.047 mM. This value is much lower than the values of Michaelis Menten reported in the literature (eg 315 mM [86]). The low of this constant indicates the high activity of Cyt c in this biosensor.

4-6 Reproducibility, reproducibility, and viability of a Cyt-based biosensor

The reproducibility of the modified electrode current response was evaluated for 5 times 1 mM NO measurement. The result for the relative standard deviation (RSD) of these results was 4.38%. This number indicates the appropriate reproducibility of the electrode.

The reproducibility of the current response of 5 electrodes made similar in 0.1 M PBS buffer (7pH) in the presence of 1 mM NO was investigated. The RSD of this test was 3.4%. This number implies high reproducibility of the fabricated electrode.

To check the stability of the fabricated electrode, the electrode was stored and tested for 4 weeks in 0.1 M PBS (7pH) buffer at 4 ° C. After two weeks of electrode storage under these conditions, little change was observed in the electrode current response to NO. After 4 weeks, the flow response changed only 3.7% compared to baseline. This high stability is due to the high biocompatibility of CS-MPA-AuNPs nanocomposites and the strong binding of AuNPs and CS.

4-7 Selectivity of Biosensor Based on Cyt c

As mentioned earlier, nephion is a material highly permeable to NO gas, while preventing the penetration of interfering anions such as nitrite (NO2−) and ascorbic acid (AA) to the surface of the modified electrode. The selectivity of the modified electrode to detect NO was evaluated by sequential injection of 15 μM NO, 150 μM NO2 NO and 150 μM AA in 0.1 M PBS buffer (7pH). The results of this experiment can be seen in Figure 25.

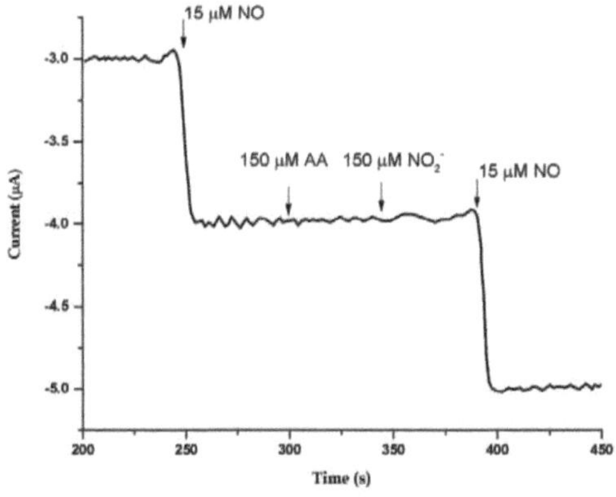

Figure (25): Amplometric response diagram of Nafion / Cyt c / CS-MPA-AuNPs / SAMs-Au electrode in 0.1 M PBS buffer at -0.7 V potential (relative to reference electrode) with sequential injection NO 15 μM, NO2 NO 150 μM and AA 150 μM

It is quite clear that the injection of interfering substances (NO2− and AA) had no effect on the flow response. In fact, the nephrons immobilized on the Cyt c / MPa-AuNPs-CS / SAMs-Au electrode significantly prevented these compounds from interfering with NO detection. Therefore, it can be said that this biosensor is selective for detecting NO in the presence of confounding compounds.

4-8 Effect of Nephion Value on Biosensor Selectivity Based on Cyt c

The role of Nephi in preventing nuisance intervention was discussed in detail in the previous section. Obviously, as nephron thickness increases, the response time increases. Therefore, optimization of nephron thickness is essential to achieve the best selectivity and response time. In Figure (26), the effect of the amount of nephrite coated on the electrode surface on the electrode response against NO and NO2− is shown.

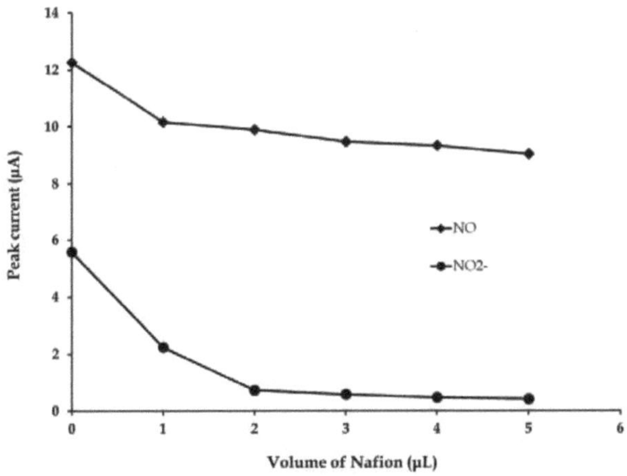

Figure (26): Influence of the Nefion value on the Nafion / Cyt c / CS-MPA-AuNPs / SAMs-Au electrode current response in 0.1 M PBS buffer at -0.7 V potential (relative to the reference electrode) compared to 75 μM NO2 و and 50 μM NO2

As can be seen, after adding 1 μL of 0.5% nephion 0.5% on the surface of the Cyt c / CS-MPA-AuNPs / SAMs-Au electrode the NO current response decreased to 83% of the initial response value, while the NO2− current response to The initial response rate decreased by 40%. Therefore, the selectivity of NO increased with respect to NO2−. When 2 microliters or more of nephion was coated on the electrode surface, the amount of current response for NO decreased slowly. On the other hand, with the addition of 2 μL of nephion, the NO2− flow response was greatly reduced, but in larger volumes of nephion, this decrease was much less. Therefore, it can be concluded that the optimum amount of Nephion 0.5% for the best selectivity of the modified electrode for NO is 2 μL.

4-9 Graphene-based NO biosensor

4-9-1 Spectroscopy of PANI-rGO and AuNPs-PANI-rGO nanocomposites

UV-vis test was used to evaluate the formation of PANI-rGO and AuNPs-PANI-rGO nanocomposites. The peak of absorption peak for rGO-PANI (dot curve) compared to GO (dot curve) was 230 nm from 270 nm. This redshift indicates the electron bonding on the graphene plates. The most likely reductive mechanism is that the released electrons were transported in the aniline polymerization process in the vicinity of oxygenated GO components and reduced GO to graphene.

4-9-2 Investigation of electrode morphology Nafion / AuNPs-PANI-ErGO / Au

FESEM test was used to evaluate the surface of the modified ErGO / Au, PANI-ErGO / Au and AuNPs-PANI-ErGO / Au electrodes. It can be concluded that by reducing the potential to negative values, GO is well electrochemically recovered to graphene. It can be seen that the AuNPs-PANI-ErGO nanocomposite has a high surface area which results in improved electron transfer at the electrode surface.

4-9-3 Effect of graphene-based electrode modification on voltammetric response

Cyclic voltammetry (CV) was used to investigate the effect of electrode modification on sensor performance. In the case of the PANI-ErGO / Au electrode (dashed curve), the peak anodic current was lower than that of the ErGO / Au electrode (point curve), due to the strong confinement of graphene active sites in the PANI macromolecular structure Electron transfer is blocked [59]. In addition, the peak anodic potential was also slightly increased (from 0.3 to 0.35 V) due to the lower electrical conductivity of the PANI-ErGO electrode [60]. The highest anodic peak current is due to the AuNPs-PANI-ErGO / Au electrode (filled curve) resulting from the high surface area and excellent electron transfer properties of AuNPs. In fact, the AuNPs-PANI-ErGO / Au electrode showed the highest conductivity, which facilitated the arrival of the redox couple [Fe (CN) 6] 3− / 4− to the electrode surface.

4-9-4 Investigation of electrochemical behavior of graphene-based sensor against NO oxidation

The CV test was performed in the potential range of 0 to V 1+ (scanning rate 100 mV / s) in the absence and presence of different NO concentrations. The modified electrode in the presence of NO has a peak oxidation potential of +8.8 V, whereas in the absence of NO it does not show any peak. This is because the PBS solution in this range of potentials (0 to V 1) and 7pH has no redox peak and the peak created in the presence of NO is due to NO oxidation. The significant increase in the peak oxidation size indicates the electrocatalytic NO oxidation process compared to its absence. Based on the results, the amperometric tests should be performed at a difference of constant potential of +8.8 V with respect to the reference electrode.

4-9-5 Investigation of the effect of pH on graphene-based biosensor performance

To investigate the effect of pH on the current response of the sensor made in the detection of NO50 micromolar, an amprometric test was performed at a potential of +8.8V at pH range 5 to 8.

4-9-6 Graphene-based sensor amperometric response to NO

After determining the optimum operating conditions, a calibration diagram must be obtained to apply the built-in biosensor. To do this, an amperometric test must first be performed under optimum operating conditions. Amperometric response of biosensor was recorded by injection of different concentrations of NO in 0.1 M PBS (7pH) buffered oxygen at 100 mV / s scanning under stirred gentle conditions. The flow induced by NO oxidation increased sharply to the maximum steady state value, reaching 95% of this maximum within 4 seconds. So the response time of this biosensor is 4 seconds. Such a short response time is due to the high electrical conductivity of the AuNPs-PANI-ErGO nanocomposite. It is observed that as the concentration of injectable NO increases, the flow response also increases.

4-9-7 Reproducibility, reproducibility and stability of the graphene-based sensor

The reproducibility of the modified electrode current response was evaluated for 5 times 1 mM NO measurement. The result for the relative standard deviation (RSD) of these results was 2.1%. This number indicates the appropriate reproducibility of the electrode.

The reproducibility of the current response of 5 electrodes made similar in 0.1 M PBS buffer (7pH) in the presence of 1 mM NO was investigated. The RSD of this test was 3.2%. This number implies the high reproducibility of the fabricated electrode.

To check the stability of the fabricated electrode, the electrode was stored and tested for 4 weeks in 0.1 M PBS (7pH) buffer at 4 ° C. After 4 weeks the flow response changed only

4.2% compared to baseline. This high stability is due to the high stability and strong binding of AuNPs, ErGO and PANI.

4-9-8 Graphene-based sensor selectivity

In the preceding sections, the role of Nephi in preventing intrusive substances is discussed in detail. Graphene-based modified electrode selectivity for NO detection by sequential injection of 10 μM NO (3 consecutive injections) 500 μM AA, 500 μM NO2, 500 μM H2O2 and again 10 μM NO (2 consecutive injections) at site. 0.1 M PBS (7pH) was evaluated. It is clear that the injection of the interfering substances (NO2−, AA and H2O2) had no effect on the flow response. In fact, the immobilized immersion on the AuNPs-PANI-ErGO / Au electrode significantly prevented these compounds from interfering with NO detection. Therefore, it can be said that this biosensor is selective for detecting NO in the presence of confounding compounds.

4-9-9 Effect of Nephion Value on the Selectivity of Graphene-Based Sensors

In the previous section, the necessity of optimizing the Nephi thickness was discussed. After adding 1 μL of 0.1% nephion on the surface of the AuNPs-PANI-ErGO / Au electrode, the NO response decreased to 63% of the initial response, while the NO2− response decreased to 48% of the initial response. Therefore, the selectivity of NO increased with respect to NO2−. When 2 microliters or more of nephion was coated on the electrode surface, the amount of current response for NO decreased slowly. On the other hand, with the addition of 2 μL of nephion, the NO2− flow response was greatly reduced, but in larger volumes of nephion, this decrease was much less. Therefore, it can be concluded that the optimum amount of 0.1% nephion to give the best selectivity of the modified electrode for NO is 3 μL.

4-10 Enzyme-based NO Sensor Cyt c Oxidase

4-10-1 Investigation of electrochemical behavior of Cyt c Oxidase-based biosensor against NO reduction

The CV test was performed in the potential range of +1.0 to +2.2 (scanning rate 100 mV / s) in the absence and presence of different NO concentrations. The modified electrode had a reduction peak in the presence of -0.7 V potential, whereas in the absence of NO it did not show any peak. This is because the PBS solution in this range of potentials (-1 to 0.2 V) and 7pH has no redox peak and the peaks generated in the presence of NO are due to NO reduction. The significant increase in the size of the resuscitation peak indicates the NO electrocatalytic resuscitation process compared to its absence. Based on the results, amperometric tests should be performed at a potential difference of -0.7 V relative to the reference electrode.

4-10-2 Effect of pH buffer on the performance of Cyt c Oxidase-based biosensor

To investigate the effect of pH on the biosensor current response made in the detection of NO50 micromolar, an amperometric test was performed at a potential of -0.7V in the pH range of 5 to 8.

4-10-3 Investigation of the enzyme concentration effect

To determine the optimum concentration of enzyme, modified electrode with different concentrations of enzyme (0.1 to 5 mg / ml) was made. Since the response time is below 6 seconds in this study, it is therefore important to apply a concentration of enzyme that generates both a high current response and a response time below 6 seconds.

As can be seen, the amount of flow response increased with increasing enzyme concentration from 0.1 to 5 mg / ml, from 25.24 to 29.10.25 μA. This increase in current indicates that as the concentration of enzyme increases, the amount of NO decreases per unit time resulting in a higher current at the electrode surface. On the other hand, it is observed that as the concentration of enzyme increases, the reaction time also increases. The reason for the increase in reaction time with increasing enzyme concentration is that as the amount of enzyme increases, the enzyme layer on the electrode surface thickens, resulting in a slow diffusion and mass transfer to the enzyme in the substrates. Take it. Due to the short shelf life of NO (6 sec), the response time below 6 sec is acceptable, so concentrations of 0.5 mg / ml above the enzyme are not acceptable. Of the acceptable concentrations, the concentration of 0.5 mg / ml produced the highest flow response. Therefore, it can be concluded that the optimum concentration of enzyme in NO reduction is 0.5 mg / ml.

4-10-4 Amperometric response of Cyt c Oxidase-based biosensor to NO

After determining the optimum operating conditions, a calibration diagram must be obtained to apply the built-in biosensor. To do this, an amperometric test must first be performed under optimum operating conditions. Amperometric response of biosensor was recorded by injection of different concentrations of NO in 0.1 M PBS (7pH) buffered oxygen at 100 mV / s scanning under stirred gentle conditions. The response time of this biosensor was estimated to be 4 seconds. Such a short response time is due to the high electrical conductivity of the AuNPs-PANI-ErGO nanocomposites and the prominent catalytic role of the enzyme in NO reduction. It is observed that as the concentration of injectable NO increases, the flow response also increases.

4-10-5 Reproducibility, Reproducibility and Stability of Cyt c Oxidase-Based Biosensor

The reproducibility of the modified electrode current response was evaluated for 5 times 1 mM NO measurement. The result for the relative standard deviation (RSD) of these results was 4.83%. This number indicates the appropriate reproducibility of the electrode.

The reproducibility of the current response of 5 electrodes made similar in 0.1 M PBS buffer (7pH) in the presence of 1 mM NO was investigated. The RSD of this test was 3.8%. This number implies high reproducibility of the fabricated electrode. To check the stability of the fabricated electrode, the electrode was stored and tested for 4 weeks in 0.1 M PBS (7pH) buffer at 4 ° C. After 4 weeks, the flow response changed only 5.6% compared to baseline. This high stability is due to the high biocompatibility of CS-MPA-AuNPs nanocomposites and the strong binding of AuNPs and CS.

4-10-6 Selectivity of Cyt c Oxidase-Based Biosensor

Graphene-based modified electrode selectivity for detecting NO by sequential injection of 10 μM NO (3 consecutive injections) 500 μM AA, 500 μM NO2, 500 μM H2O2 and again 10 μM NO (2 consecutive injections) at site. 0.1 M PBS (7pH) was evaluated. It is clear that the injection of the interfering substances (NO2−, AA and H2O2) had no effect on the flow response. In fact, the nephrons immobilized on the Cyt c Oxidase / AuNPs-PANI-ErGO / Au electrode significantly prevented these compounds from interfering with NO detection. Therefore, it can be said that this biosensor is selective for detecting NO in the presence of confounding compounds.

4-10-7 Effect of Nephion Value on Cyt c Oxidase-Based Biosensor Selectivity

Obviously, as the nephron thickness increases, the response time increases. Therefore, optimization of nephron thickness is essential to achieve the best selectivity and response time. After adding 1 μl of 0.5% nephion 0.5% on the surface of the Cyt c Oxidase / AuNPs-PANI-ErGO / Au electrode, the NO response decreased to 81% of the initial response, while the NO2− response decreased to 51% of the initial response. Therefore, the selectivity of NO increased with respect to NO2−. When 2 microliters or more of nephion was coated on the electrode surface, the amount of current response for NO decreased slowly. On the other hand, with the addition of 2 and 3 μl of Nephi, the NO2− flow response was greatly reduced, but in larger Nephi volumes this decrease was much less. Therefore, it can be concluded that the optimum amount of Nephion 0.5% for the best selectivity of the modified electrode for NO is 3 μL.

CHAPTER

5

Conclusion

5-1 Introduction

In this study, two biosensors based on cytochrome c protein (Cyt c) and cytochrome c oxidase (Cyt c oxidase) and a graphene-based electrochemical sensor were developed to detect nitric oxide (NO). The basis of these three sensors was based on the 3-electrode chemical cell. In all of these sensors, the gold electrode was used as the working electrode. To improve the performance of the sensors in detecting NO, the gold working electrode was modified using various materials. After fabrication of modified electrodes, their electrochemical behavior was evaluated using various electrochemical tests such as cyclic voltammetry (CV) and chronoamprometry. To achieve the best performance of the sensors, the values of some effective operating parameters such as pH, concentration of the enzyme stabilized, and the volume of nephion were also investigated. In addition, sensors performance parameters such as sensitivity, detection limit, linear range, reproducibility, stability and selectivity were investigated and reported.

5-2 Conclusion

In summary, the results for these sensors are as follows:

1. Cyt c protein-based biosensor

- Chitosan (CS), gold nanoparticles (AuNPs), mercaptopropionic acid (MPA) and Cyt c were used to modify the working electrode in this biosensor.
- Based on the UV-vis test results, the CS-MPA-AuNPs nanocomposite was well formed and provided an environmentally friendly environment for Cyt c to maintain its catalytic activity.
- Modified Nafion / Cyt c / MPA-AuNPs-CS / SAMs-Au modified working electrode based on consolidated covalent fixation, imprinting and encapsulation.

- The effect of presence of CS and AuNPs on electrode performance was evaluated by CV test. The results showed that the structure of CS-AuNPs nanocomposite can significantly reduce the gap between the Cyt c redox centers and the gold electrode. This leads to a significant increase in the effective surface area of the modified electrode to stabilize the greater amount of Cyt c and also to provide high conductivity to improve the electron transfer between the Cyt c and the gold electrode surface.

2. Graphene-based sensor

- Modifications such as polyaniline (PANI), gold nanoparticles (AuNPs) and graphene (rGO) were used to modify the working electrode.
- Based on UV-vis test results, the AuNPs-PANI-rGO nanocomposite was well formed.
- The modified Nafion / AuNPs-PANI-ErGO / Au modified working electrode was fabricated based on a combination of electrode positioning (electrolyte-addressing) and encapsulation.
- FESEM images showed that the graphene plates, the PANI porous polymer layer, and the gold nanoparticles appeared well on the electrode surface. In addition, the proper distribution of AuNPs on the surface of the modified electrode was shown.
- The effect of presence of graphene, PANI and AuNPs on electrode performance was investigated by CV test. The results indicate that the graphene active sites are strongly encapsulated in the macromolecular structure of PANI, leading to the blockage of electron transfer. The highest anodic peak current was related to the AuNPs-PANI-ErGO / Au electrode, which is due to the high surface area and excellent electron transfer properties of AuNPs.
- Voltage graphs of the modified electrode cycles in the presence of NO showed that NO oxidation occurs at a potential of +8.8 V and that as the NO concentration increased, the peak oxidation current increased.

3. Enzyme-based biosensor Cyt c Oxidase

- This biosensor was actually a modified graphene-based sensor presented in this study.
- An enzyme layer was then stained on the AuNPs-PANI-ErGO / Au electrode and then neat to obtain the Nafion / Cyt c Oxidase / AuNPs-PANI-ErGO / Au electrode.
- FESEM images showed that the enzyme is well adsorbed on the polymeric structure of the AuNPs-PANI-ErGO nanocomposite.
- Voltage graphs of the modified electrode cycles in the presence of NO showed that NO reduction occurs at a potential of -0.7 eV and also increased with increasing NO concentration at the reduction peak current.

References

[1] Arora, K., Chand, S., & Malhotra, B. D. (2006). Recent developments in bio-molecular electronics techniques for food pathogens. *Analytica Chimica Acta, 568*(1-2), 259-274.

[2] Illanes, A. (2014). Introduction: Enzyme Biocatalysis: Principles and Applications.

[3] Katchalski-Katzir, E. (1993). Immobilized enzymes—learning from past successes and failures. *Trends in biotechnology, 11*(11), 471-478.

[4] Ho, C., Robinson, A., Miller, D., & Davis, M. (2005). Overview of sensors and needs for environmental monitoring. *Sensors, 5*(1), 4-37.

[5] Nicholas, J. V., & White, D. R. (1994). *Traceable temperatures: an introduction to temperature measurement and calibration* (Vol. 8). New York: Wiley.

[6] Eggins, B. R. Biosensors. An Introduction 1996. *John Wiley&Sons Ltd.*

[7] Kress-Rogers, E., & Brimelow, C. J. (Eds.). (2001). *Instrumentation and sensors for the food industry* (Vol. 65). Woodhead Publishing.

[8] Mulchandani, P., Chen, W., & Mulchandani, A. (2001). Flow injection amperometric enzyme biosensor for direct determination of organophosphate nerve agents. *Environmental science & technology, 35*(12), 2562-2565.

[9] Bush, D. L., & Rechnitz, G. A. (1987). Monoclonal antibody biosensor for antigen monitoring. *Analytical letters, 20*(11), 1781-1790.

[10] Karlsson, R., Michaelsson, A., & Mattsson, L. (1991). Kinetic analysis of monoclonal antibody-antigen interactions with a new biosensor based analytical system. *Journal of immunological methods, 145*(1-2), 229-240.

[11] Eggins, B. R. (2008). *Chemical sensors and biosensors* (Vol. 28). John Wiley & Sons.

[12] Nguyen, B. T. T., Peh, A. E. K., Chee, C. Y. L., Fink, K., Chow, V. T., Ng, M. M., & Toh, C. S. (2012). Electrochemical impedance spectroscopy characterization of nanoporous alumina dengue virus biosensor. *Bioelectrochemistry, 88*, 15-21.

[13] Carey, W. P. (1994). Multivariate sensor arrays as industrial and environmental monitoring systems. *TrAC Trends in Analytical Chemistry, 13*(5), 210-218.

[14] Zare, H., Najafpour, G., Jahanshahi, M., Rahimnejad, M., & Rezvani, M. (2014). Enhancement of amperometric response of glucose biosensor by electrodeposition of silver nanoparticles onto chitosan-modified electrode. *Linnaeus Eco-Tech.*

[15] Allen, B. L., Kichambare, P. D., & Star, A. (2007). Carbon nanotube field-effect-transistor-based biosensors. *Advanced Materials, 19*(11), 1439-1451.

[16] Ligler, F. S., & Taitt, C. R. (Eds.). (2002). *Optical biosensors: present & future*. Gulf Professional Publishing.

[17] Situmorang, M., Hibbert, D. B., & Gooding, J. J. (2000). An Experimental Design Study of Interferences of Clinical Relevance of a Polytyramine Immobilized-Enzyme

Biosensor. *Electroanalysis: An International Journal Devoted to Fundamental and Practical Aspects of Electroanalysis*, *12*(2), 111-119.

[18] Brena, B. M., & Batista-Viera, F. (2006). Immobilization of enzymes. In *Immobilization of enzymes and cells* (pp. 15-30). Humana Press.

[19] Messing, R. (Ed.). (2012). *Immobilized enzymes for industrial reactors*. Elsevier.

[20] Scouten, W. H., Luong, J. H., & Brown, R. S. (1995). Enzyme or protein immobilization techniques for applications in biosensor design. *Trends in biotechnology*, *13*(5), 178-185.

[21] Gupta, A., Avci, P., Sadasivam, M., Chandran, R., Parizotto, N., Vecchio, D., ... & Hamblin, M. R. (2013). Shining light on nanotechnology to help repair and regeneration. *Biotechnology advances*, *31*(5), 607-631.

[22] Wang, J. (2005). Nanomaterial-based electrochemical biosensors. *Analyst*, *130*(4), 421-426.

[23] Fortina, P., Kricka, L. J., Surrey, S., & Grodzinski, P. (2005). Nanobiotechnology: the promise and reality of new approaches to molecular recognition. *TRENDS in Biotechnology*, *23*(4), 168-173.

[24] Mu, C., Zhao, Q., Xu, D., Zhuang, Q., & Shao, Y. (2007). Silicon nanotube array/gold electrode for direct electrochemistry of cytochrome c. *The Journal of Physical Chemistry B*, *111*(6), 1491-1495.

[25] Hasanzadeh, M., Shadjou, N., Eskandani, M., & de la Guardia, M. (2012). Mesoporous silica-based materials for use in electrochemical enzyme nanobiosensors. *TrAC Trends in Analytical Chemistry*, *40*, 106-118.

[26] Suni, I. I. (2008). Impedance methods for electrochemical sensors using nanomaterials. *TrAC Trends in Analytical Chemistry*, *27*(7), 604-611.

[27] Iijima, S., & Ichihashi, T. (1993). Single-shell carbon nanotubes of 1-nm diameter. *nature*, *363*(6430), 603.

[28] Bidan, G. (1992). Electroconducting conjugated polymers: new sensitive matrices to build up chemical or electrochemical sensors. A review. *Sensors and Actuators B: Chemical*, *6*(1-3), 45-56.

[29] Lange, U., Roznyatovskaya, N. V., & Mirsky, V. M. (2008). Conducting polymers in chemical sensors and arrays. *Analytica chimica acta*, *614*(1), 1-26.

[30] Palaniappan, S., & John, A. (2008). Polyaniline materials by emulsion polymerization pathway. *Progress in polymer science*, *33*(7), 732-758.

[31] Yoon, H., & Jang, J. (2009). Conducting-polymer nanomaterials for high-performance sensor applications: issues and challenges. *Advanced Functional Materials*, *19*(10), 1567-1576.

[32] Katz, E., Willner, I., & Wang, J. (2004). Electroanalytical and bioelectroanalytical systems based on metal and semiconductor nanoparticles. *Electroanalysis: An International Journal Devoted to Fundamental and Practical Aspects of Electroanalysis*, *16*(1-2), 19-44.

[33] Cao, X., Ye, Y., & Liu, S. (2011). Gold nanoparticle-based signal amplification for biosensing. *Analytical biochemistry*, *417*(1), 1-16.

[34] Guo, S., & Wang, E. (2007). Synthesis and electrochemical applications of gold nanoparticles. *Analytica chimica acta*, *598*(2), 181-192.

[35] Gupta, A. K., & Gupta, M. (2005). Synthesis and surface engineering of iron oxide nanoparticles for biomedical applications. *biomaterials*, *26*(18), 3995-4021.

[36] Jiang, Y., Zhang, Q., Li, F., & Niu, L. (2012). Glucose oxidase and graphene bionanocomposite bridged by ionic liquid unit for glucose biosensing application. *Sensors and Actuators B: Chemical*, *161*(1), 728-733.

[37] José-Yacamán, M., Rendón, L., Arenas, J., & Puche, M. C. S. (1996). Maya blue paint: an ancient nanostructured material. *Science*, *273*(5272), 223-225.

[38] Andrade, C. A., Oliveira, M. D., De Melo, C. P., Coelho, L. C., Correia, M. T., Nogueira, M. L., ... & Zeng, X. (2011). Diagnosis of dengue infection using a modified gold electrode with hybrid organic–inorganic nanocomposite and Bauhinia monandra lectin. *Journal of colloid and interface science*, *362*(2), 517-523.

[39] Kuznetsov, K. A., Laptinskaya, T. V., Mamaeva, Y. B., & Shapiro, B. I. (2006). SELF-ASSEMBLY TECHNIQUE. *Electronic Processes in Organic Materials*, *2006*, 95.

[40] Bian, J. C., Chen, Z. D., Li, Z., Yang, F., He, H. Y., Wang, J., ... & Han, G. R. (2012). Electrodeposition of hierarchical Ag nanostructures on ITO glass for reproducible and sensitive SERS application. *Applied Surface Science*, *258*(17), 6632-6636.

[41] Qu, F., Sun, H., Zhang, Y., Lu, H., & Yang, M. (2012). Electrochemically deposited Pd nanorod array/sol–gel silica thin film for the fabrication of electrochemical sensors. *Sensors and Actuators B: Chemical*, *166*, 837-841.

[42] Kobayashi, Y., Salgueirino-Maceira, V., & Liz-Marzán, L. M. (2001). Deposition of silver nanoparticles on silica spheres by pretreatment steps in electroless plating. *Chemistry of Materials*, *13*(5), 1630-1633.

[43] Kaisheva, A., Iliev, I., Christov, S., & Kazareva, R. (1997). Electrochemical gas biosensor for phenol. *Sensors and Actuators B: Chemical*, *44*(1-3), 571-577.

[44] Dennison, M. J., Hall, J. M., & Turner, A. P. (1995). Gas-phase microbiosensor for monitoring phenol vapor at ppb levels. *Analytical Chemistry*, *67*(21), 3922-3927.

[45] Mitsubayashi, K., Yokoyama, K., Takeuchi, T., & Karube, I. (1994). Gas-phase biosensor for ethanol. *Analytical Chemistry*, *66*(20), 3297-3302.

[46] Hämmerle, M., Hall, E. A., Cade, N., & Hodgins, D. (1996). Electrochemical enzyme sensor for formaldehyde operating in the gas phase. *Biosensors and Bioelectronics*, *11*(3), 239-246.

[47] Fauzia, V., Imawan, C., Narayani, N. M. M. S., & Putri, A. E. (2018). A localized surface plasmon resonance enhanced dye-based biosensor for formaldehyde detection. *Sensors and Actuators B: Chemical*, *257*, 1128-1133.

[48] He, Q., Zheng, D., & Hu, S. (2009). Development and application of a nano-alumina based nitric oxide sensor. *Microchimica Acta*, *164*(3-4), 459-464.

[49] Deng, X., Wang, F., & Chen, Z. (2010). A novel electrochemical sensor based on nano-structured film electrode for monitoring nitric oxide in living tissues. *Talanta*, *82*(4), 1218-1224.

[50] Bai, R. G., Muthoosamy, K., Zhou, M., Ashokkumar, M., Huang, N. M., & Manickam, S. (2017). Sonochemical and sustainable synthesis of graphene-gold (G-Au) nanocomposites for

enzymeless and selective electrochemical detection of nitric oxide. *Biosensors and Bioelectronics*, *87*, 622-629.

[51] Kelm, M., Feelisch, M., Deuen, A., Schrader, J., & Strauer, B. E. (1991). The role of nitric oxide in the control of coronary vascular tone in relation to partial oxygen pressure, perfusion pressure, and flow. *Journal of cardiovascular pharmacology*, *17*, S95-S99.

[52] Pashai, E., Darzi, G. N., Jahanshahi, M., Yazdian, F., & Rahimnejad, M. (2018). An electrochemical nitric oxide biosensor based on immobilized cytochrome c on a chitosan-gold nanocomposite modified gold electrode. *International journal of biological macromolecules*, *108*, 250-258.

[53] Sekiguchi, A., Kinjo, R., & Ichinohe, M. (2004). A stable compound containing a silicon-silicon triple bond. *Science*, *305*(5691), 1755-1757.

[54] Liu, Y., Wang, M., Zhao, F., Xu, Z., & Dong, S. (2005). The direct electron transfer of glucose oxidase and glucose biosensor based on carbon nanotubes/chitosan matrix. Biosensors and Bioelectronics, 21(6), 984-988.

[55] Li, C. M., Zang, J., Zhan, D., Chen, W., Sun, C. Q., Teo, A. L., ... & Moochhala, S. M. (2006). Electrochemical detection of nitric oxide on a SWCNT/RTIL composite gel microelectrode. *Electroanalysis: An International Journal Devoted to Fundamental and Practical Aspects of Electroanalysis*, *18*(7), 713-718.

[56] Grieshaber, D., MacKenzie, R., Vörös, J., & Reimhult, E. (2008). Electrochemical biosensors-sensor principles and architectures. Sensors, 8(3), 1400-1458.

[57] Xu, J., Shang, F., Luong, J. H., Razeeb, K. M., & Glennon, J. D. (2010). Direct electrochemistry of horseradish peroxidase immobilized on a monolayer modified nanowire array electrode. Biosensors and Bioelectronics, 25(6), 1313-1318.

[58] Lu, Q., Hu, S., Pang, D., & He, Z. (2005). Direct electrochemistry and electrocatalysis with hemoglobin in water-soluble quantum dots film on glassy carbon electrode. *Chemical Communications*, (20), 2584-2585.

[59] Srivastava, S., Kumar, V., Ali, M. A., Solanki, P. R., Srivastava, A., Sumana, G., ... & Malhotra, B. D. (2013). Electrophoretically deposited reduced graphene oxide platform for food toxin detection. *Nanoscale*, *5*(7), 3043-3051.

[60] Li, Q., Cheng, K., Weng, W., Du, P., & Han, G. (2012). Highly sensitive hydrogen peroxide biosensors based on TiO 2 nanodots/ITO electrodes. *Journal of Materials Chemistry*, *22*(18), 9019-9026.

MIX
Papier aus verantwortungsvollen Quellen
Paper from responsible sources
FSC® C105338

Printed by Books on Demand GmbH, Norderstedt / Germany